S0-BWQ-322

In the Bight

IN THE BIGHT

The BC Forest Industry Today

Ken Drushka

HARBOUR PUBLISHING

Published by
HARBOUR PUBLISHING
P.O. Box 219
Madeira Park, BC Canada
V0N 2H0

Cover design by Martin Nichols, Lionheart Graphics
Edited by Michael Maser

Printed and bound in Canada

Harbour Publishing acknowledges the financial support of the Government of Canada through the Book Publishing Industry Development Program (BPIDP) and the Canada Council for the Arts, and the Province of British Columbia through the British Columbia Arts Council, for its publishing activities.

Canadian Cataloguing in Publication Data

Drushka, Ken.
In the bight

Includes bibliographical references and index.
ISBN 1-55017-161-3

1. Forests and forestry—British Columbia. 2. Forest policy—British Columbia. 3. Forest management—British Columbia. I. Title.
SD146.B7D78 1999 333.75'09711 C99-910760-1

Contents

Introduction

The British Columbia forest industry is in trouble. In the face of the largest lumber sales in history, the BC industry spent much of the late 1990s awash in red ink. Mills closed, some of them permanently, workers lost their jobs, and an atmosphere of pessimism has prevailed throughout the industry, in spite of occasional surges of activity in forest product markets.

The irony of the current situation is that the opportunities available to the forest industry are enormous—more than ever before in its 130-year history. Markets for renewable resource products, including timber, are strong and growing. An expanding world population needs and wants lumber, plywood, pulp, paper and other forest products that BC could, if we were properly organized, provide in increased quantities far into the future.

The means to build a foundation for a prosperous economy on the stewardship of natural resources is well within our grasp. We have the land base for it; we have skills and expertise to build upon; and, for the first time in our history, we have public interest and support for a profoundly different approach to the use we make of BC's forests.

Why, then, is the BC industry faltering? Why, instead of embarking on a prosperous future, is the industry wallowing in despair, anticipating even tougher times ahead? What is missing?

What we are lacking is a vision and a strategy to realize that prosperous future.

This book does not presume to provide a panacea. It looks at the development of forest policy in BC in an attempt to explain why the industry had found itself at an impasse by the late twentieth century. It describes the consequences of some of these policies and offers a few alternatives and approaches worth discussing as government, industry, labour and other British Columbians search for solutions.

In some ways this book is a sequel to or update of an earlier book of mine, *Stumped*, published in 1983. That book cheered some readers and infuriated others, but whatever the reaction, the book is fifteen years old: policy, industrial realities, public debate and the forest itself have changed since then. At the urging of others I compiled this new material, and I hope it will contribute more up-to-date information and analysis to discussions about the future of the forest industry in BC.

When I wrote *Stumped* I was coming to the end of more than fifteen years of working in the forest industry, although I did not know it at the time. Most of this time had been spent in or around Campbell River, on northern Vancouver Island. My perspective then was that of a woods worker and gyppo sawmill operator, with a focus on the BC coast. Since then I have lived in Vancouver and written books and articles about the rest of the forest industry. I have had the opportunity to spend considerable time studying the history of the industry, as well as obtaining a first-hand look at how it operates in the rest of BC, other parts of Canada, the United States and Finland. And it has been my good fortune to meet a lot of people passionately involved with forests in various capacities.

Looking back I realize that some of the opinions expressed and judgements passed in *Stumped* have stood the test of time and some have not. In the mid-1980s, for example, I was convinced large forest corporations were, by their very existence, a primary cause of the industry's problems. Maybe the onset of advanced middle age has something to do with it, but having seen how responsibly big companies perform in other jurisdictions, I am reluctant to support unequivocally such a conclusion today. I now suggest it is the context in which these companies have had to operate in BC that creates most of the problems associated with them.

I am still able to spend time in forests and, occasionally, to fire up my sawmill. I no longer earn my living this way, but it brings me back to earth. More than ever I find that the best place to think and talk about forest policy is in a forest. A few years ago I ended up in yet another argument with the late Mark Waering, then a forester with the Western Canada Wilderness Committee. We had encountered each other a number of times, usually on a public platform or at a media event, and invariably ended up in strong disagreement.

"Mark," I said on this occasion, "I refuse to discuss this or anything else with you if we are inside a building. If you want to talk about it any more, meet me at the UBC Endowment Lands forest tomorrow morning."

The next day we spent several hours walking through the forest, discussing how it could be managed if it were a working forest, and what we would do with it if the decisions were ours. In the end we found we agreed almost completely. I learned that day that practising good forestry is not very difficult; where the problems arise is in talking about how to practise forestry. I heartily recommend this on-the-ground method to politicians, corporate executives, bureaucrats, environmentalists, foresters and anyone else who is in any position to learn about, discuss or develop forest policy. It simplifies things enormously.

As is always the case, I owe a huge debt to the many people who have attempted, not always successfully, to enlighten me on some issue or other relating to forests. They are too numerous to list here and some of them may prefer not to have it known that they talk to me. But I would like to say thank you to a few extraordinary people whose concern for the future of BC's forests is paramount: Jim Collins, Peter Pearse, Robin Clark and Ray Travers. You have done me the favour, at one time or another, of telling me I was wrong and taking the time and effort to set me straight. There are scores of others who work in the forest industry, toil unappreciated in the Ministry of Forests, or teach at educational institutions throughout the province who have contributed enormously to my understanding. One of them is my wife, who has become extremely adept at changing the subject when I try to entice her into dinnertime discussion about the latest forest policy issue. To all of you, thanks, and if I didn't get it right it wasn't because you didn't try.

The Forests of BC 1

British Columbia is part of an ancient landform that first emerged from beneath the Pacific Ocean almost 200 million years ago, along with southern Alaska, Washington state, most of Oregon and a bit of northern California. Forced upward by shifting continental plates, the land created was rugged, with a series of north–south mountain ranges rising between the Pacific shore and the plains to the east.

The geographic features that most influence the region are the Rocky Mountains to the east and the Pacific Ocean to the west. As moist air from the Pacific moves east and rises to pass over a succession of mountain ranges, it loses its moisture in the form of rain and snow, creating a great network of rivers and lakes draining back into the Pacific. The entire area is dominated by three great river systems, the Columbia, the Fraser and the Skeena. It embraces a great inland sea, the Strait of Georgia and Puget Sound. With the exception of a dry interior desert extending from south-central BC to California, it is a land of cascading waters, producing prolific and diverse forest ecosystems that are among the most biologically productive in the world.

The land was given its basic physiological form—its mountain ranges and ocean depths—by primeval forces we can barely imagine. This expanse of solid bedrock was then scoured and polished by a succession of glaciers, beginning some two million years ago.

Between these glacial periods, for terms of 50,000 or 60,000 years, dinosaurs and other creatures lived in forests of tree species some of which no longer exist, and some of which now grow only in other parts of the world. For example, fossils reveal that about fifty million years ago, the dominant tree species in the region was the metasequoia, or dawn redwood, which covered much of the province. Today this species grows wild only in a remote area of China.

The last ice age to affect BC, known as the Fraser Glaciation, began about 30,000 years ago. For the next 25,000 years, with temperatures hovering near or below freezing, the moist Pacific air unloaded a steady fall of snow that covered the mountains and northern regions, then the lower elevations and the southern and coastal areas. The snow fell for another 10,000 years, compacting into ice several thousand feet thick. As it gained density and weight, the ice flowed down the ancient watersheds to the ocean, scouring the land and grinding the bedrock into finer particles, which it then deposited along its course.

Slowly the flora and fauna retreated before the advancing ice. Global living space contracted dramatically, and trees that were once found in what is now northern BC began to grow in California. Within BC, only a few peninsulas and offshore islands were untouched by the ice. The Brooks Peninsula on the west coast of Vancouver Island was one such refuge, and the Queen Charlotte Islands were another.

And then, as inexplicably as it had begun, the ice age ended. About 15,000 years ago temperatures began to rise, snow changed to rain and the ice began to melt. The particles of bedrock that had been scraped loose were washed into valley bottoms, leaving the mountain slopes bare. The land emerged sterile and barren from the ice. For almost 5,000 years the ice retreated to the relatively stable area it occupies today in the Coast and St. Elias mountain ranges.

From their small refuges, the plant and animal species that had escaped the ice began to reclaim the landscape, migrating along the coastal margin and up the river valleys. Smaller plants such as grasses and shrubs established themselves first, their seeds borne on the wind or carried onto the scoured landscape by birds and animals.

Trees advanced much more slowly, at an average rate of about 300 metres a year. The pioneering species like red alder and lodgepole pine—those that are still the first to grow after a fire—spread most quickly. Lodgepole pine, having found refuge on the Brooks Peninsula, appeared in the mountains of the Lower Mainland about 13,000 years ago, in much the same way it occupied the logged-off, burned-over land west of Qualicum Beach in the latter half of this century. From the same refuge, mountain hemlock gradually spread throughout the coastal mountains of Vancouver Island and the mainland.

Douglas fir, a less suitable pioneering species because it requires at least a thin covering of humus soil and full exposure to the sun, edged its way in from California about 3,000 years later. Initially it established itself where lightning touched off fires in the pioneer lodgepole pine stands. The coastal lands around the Strait of Georgia and the river valleys of the southern Interior were particularly suited to Douglas fir and it thrived there, dominating the established pioneers and growing in magnificent stands.

Gradually, over the centuries, other species began to move into the Douglas fir zone, occupying the wetter sites and prevailing in the outer coastal margins. About 5,000 years after Douglas fir arrived, western red cedar and western hemlock became established throughout the coastal region and the southeastern Interior. Western white pine and bigleaf maple moved north. Sitka spruce migrated from the Queen Charlotte Islands and the Brooks Peninsula, along with yellow cedar, Amabilis fir and a variety of minor species.

The central Interior of the province was a meeting ground of dispersed tree species. Engelmann spruce moved north from its refuge in the American southwest, while white spruce, its close relative, appears to have survived the ice age in a northern refuge. In the higher, colder and drier Chilcotin plateau, on the eastern flank of the Coast Mountains in whose rain shadow it lies, lodgepole pine arrived first and has dominated ever since.

Farther north, in the top one-third of the province, from the Alaska border to the Rocky Mountains, lies a vast area that has been only marginally repopulated by trees. The ice left this land reluctantly and lingers still in some places, threatening to return during every long, cold winter. Much of this area is tundra, a semi-barren landscape in which scrub species and stunted versions of southern species grow. White spruce may have found refuge here, if not farther north. The intrepid lodgepole pine has reached this area, along with subalpine fir, willow and birch.

Birch may have come to the province from the east, over the Rockies, when the continental ice sheets melted. There the land was reclaimed by the great northern boreal forest that stretches across northern Canada from Alaska to Newfoundland. The trees here are spruces, which may have waited out the ice in northern refuges, and aspen, a deciduous species that followed the retreating ice north across the prairies.

While the land itself may be old, what grows and lives upon it is very young: these forests are the dominant ecological attribute of the region and contain some of the oldest, largest and densest stands of trees in the world. In fact, the common characteristic of BC's forests is their youthfulness. Some people speak easily of the province's "ancient" forests, but relatively speaking they are very young—only three or four thousand years old.

What is unique about BC's forests is the diversity of the province's forest ecosystems. In fact there is probably no known place on earth of comparable size that contains so many different types of forests. Geologists identify five broadly defined physiographic regions in the province: the Coast Mountains and islands, the Interior plateau, the northern and central plateaus and mountains, the great plains, and the Columbia Mountains and southern Rockies. The configuration of the largely mountainous landform and maritime weather conditions produces a wide mix of climates ranging from Mediterranean on the southwest coast to semi-desert conditions in the southern Interior to subarctic at higher mountain elevations. It has also produced, over time, a diverse amalgam of soil types comprising nine major groupings under the Canadian system of soil classification.

Thanks to these unique combinations of landform, climate and soil, a wide variety of distinct biological communities have evolved in BC. There are extreme differences, for instance, between the near-desert habitat around Osoyoos, the soggy conditions of Ocean Falls and the frigid atmosphere of the Tatshenshini River valley. There are also local diversities: at the head of Knight Inlet, maritime conditions are found cheek by jowl with Interior influences, and Interior-dwelling moose gaze out over the salt water. Opposite-facing slopes of the same mountain can support separate and distinct biological communities.

To work with this diversity, scientists have devised a unique classification system that divides the province into fourteen biogeoclimatic zones. The system was conceived by Vladimir Krajina, a European botanist who came to the University of British Columbia seeking refuge from the glacial chill of totalitarianism. Appalled at the mistreatment of forests in BC, he set out to find ways to better understand and appreciate them. His system, devised in the 1960s, became the foundation of forest ecology in BC.

Krajina's method made it possible, for the first time, to communicate an understanding of the province's forests that went far beyond a mere description of the volume and types of timber they contained. The system describes the particular combinations of life forms, physical attributes and climatic conditions found in each location. Among its many useful functions, the system allows us to identify how much of each forest type grows in each zone.

BIOGEOCLIMATIC ZONES
(Areas and Percentages)

Zone	**Area (million hectares)**	**% of province**
Alpine Tundra	17.5	18
Spruce–Willow–Birch	7.8	8
Boreal White & Black Spruce	15.1	16

Sub-Boreal Pine–Spruce	2.4	3
Sub-Boreal Spruce	9.9	10
Mountain Hemlock	4.1	4
Engelmann Spruce–Subalpine Fir	13.3	14
Montane Spruce	2.6	3
Bunchgrass	.3	0.3
Ponderosa Pine	.3	0.3
Interior Douglas fir	4.3	5
Coastal Douglas fir	.2	0.2
Interior Cedar–Hemlock	5.0	5
Coastal Western Hemlock	10.6	11

Source: *Forest, Range & Recreation Resource Analysis* 1994 (Victoria: Ministry of Forests), pp. 28–33.

HOW MUCH FOREST LAND THERE IS AND WHO CONTROLS IT

	Forest Land	Non-forest Land	Total	Percent of BC
	hectares			%
Public Forest Land				
Timber Supply Areas	39.5	35.5	75.0	79.0
Tree Farm Licences	3.7	3.2	6.9	7.3
Provincial/Federal Park	2.0	3.8	5.8	6.1
Private Forest Land	2.7	0.0	2.7	2.8
Subtotal Forest	47.9	42.5	90.4	95.2%
Agriculture, Urban etc.	0.0	4.2	4.2	4.4
Total Land	47.9	46.7	94.6 million	
Percentage of BC	50.6%	49.4%		100%

Source: Ministry of Forests, *Park Data Book*

An examination of these figures gives rise to some sobering facts. For instance, the coastal forest industry, upon which the economy of the province largely depended until after World War II, was based for fifty years on the coastal Douglas fir forest—which covered only two-tenths of 1 percent of the province. And the most common coastal species is western hemlock, commercially ignored until after World War II and still not highly prized.

Even in Krajina's system there are few distinct boundaries between different types of forests. They blend and merge and their borders overlap. Identifying what is forest and what is not can be equally imprecise. While moving up a mountain, when does the forest end—when the trees are smaller than a certain size? Is a slope of stunted knee-high fir and hemlock part of the forest? Are the stands of widely scattered pines in Chilcotin ranchlands forest or rangelands? Answers to these questions help determine how much of the forest can be used by whom, and for what purposes.

Measuring the size of and determining the nature of the BC forest is a challenge. BC contains 95.2 million hectares of land, including every square centimetre of ground, rock, ice and fresh water in the province. More than half of this area, about 57 million hectares, is covered by forest of one kind or another, much of it unsuitable for timber harvesting operations. The rest consists of unforested mountaintops, glaciers, tundra, water, open grasslands, hydro reservoirs and transmission corridors, urban areas and other areas devoid of trees.

Of the area classified as forest in 1994, about 8 million hectares was in parks—600,000 hectares in national parks and the rest provincially owned. A further subclassification identifies forest land that is commercially productive, that is, capable of growing merchantable timber for use by the forest industry. This figure is arrived at by subtracting unproductive lands including alpine forests, forests growing on swamps and muskeg, or on steep, rocky and broken sites, as well as areas not available for commercial use, such as areas being studied as possible habitat for threatened species. What was left as productive forest until the mid-1990s, when more parks and other protected areas were set aside, amounted to about 49 million hectares. By 1998 the government, as part of its stated intention to protect 12 percent of the provincial land base from any commercial use, had assigned protected status to an additional 2 million hectares, bringing the total to 10.6 percent of the province.

During this same period the government passed the Forest Land Reserve Act to provide the forest industry with a stable land base by designating commercial forest land. The intention was that by the year 2000, all forest land in the province would be classified and all commercial forest land placed in the Reserve.

The BC government is granted control of forests within its boundaries by the Canadian constitution. For administrative purposes, the

FOREST REGIONS OF BC

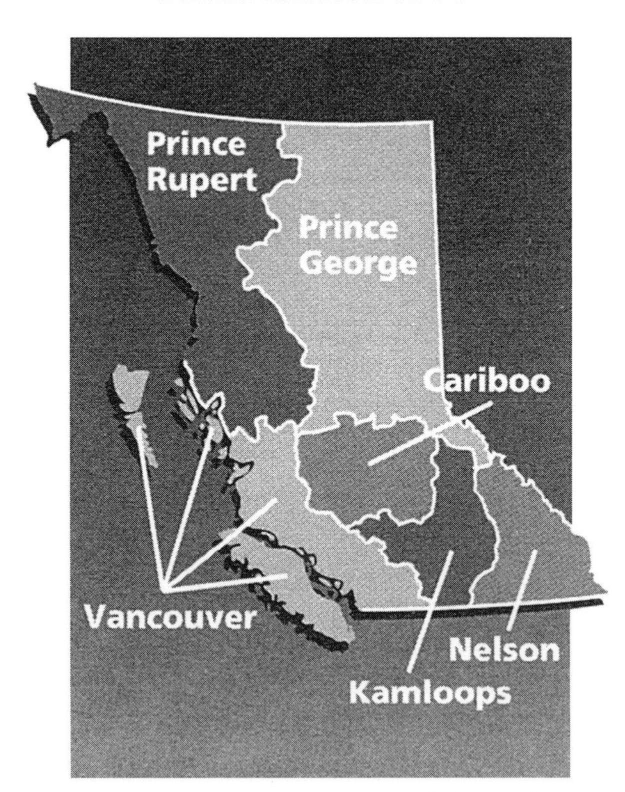

Source: *Forest, Range & Recreation Resource Analysis* 1994 (Victoria: Ministry of Forests).

province's forest lands are divided into six forest regions, and each region is further divided into several districts, all under the jurisdiction of the Ministry of Forests. Most of the ministry's functions are performed in district offices, but much of the statistical and inventory data is compiled on a regional basis (see Appendix, Table 1A).

The government also categorizes land on the basis of tenure, which defines the way in which rights to forest land are held. As in most Canadian provinces, almost all forest land in BC is owned by government. About 2.5 million hectares are privately owned, although estimates vary. About 1 million hectares of this is industrial forest land, four-fifths of it owned by large forest companies. The remainder is held in small parcels, most of it unmanaged land on which harvesting is not regulated. About 48 percent of private forest land is in the Prince George forest region, and while the east coast of Vancouver Island contains a concentration of the most productive forest land in BC, only 4 percent of the province's private forest land is found in the Vancouver forest region. Federally controlled land in the province consists of about 340,000 hectares of Indian reserves, 600,000 hectares of national parks, and less than 100,000 hectares of military reserves and other blocks. Much of the federal land is forested, although relatively little is available for commercial use.

Provincially owned forest land is organized into two categories. Tree Farm Licences (TFLs) are relatively large areas leased to private firms and municipal bodies for the purpose of harvesting timber and growing future timber crops. The licence holders manage the land under terms and conditions prescribed by the government. The second and largest portion of BC-owned forest land is found in Timber Supply Areas (TSAs), which are managed by the Ministry of Forests in close collaboration with the forest industry. Timber from TSAs is sold to industry under a variety of leases and licences on an annual volume basis. A very small portion of TSA lands is leased to private operators in the form of Woodlot Licences or small parcels of forest land managed under similar terms as TFLs (see Appendix, Tables 1B and 1C).

DISTRIBUTION OF TREE FARM LICENCES IN BC

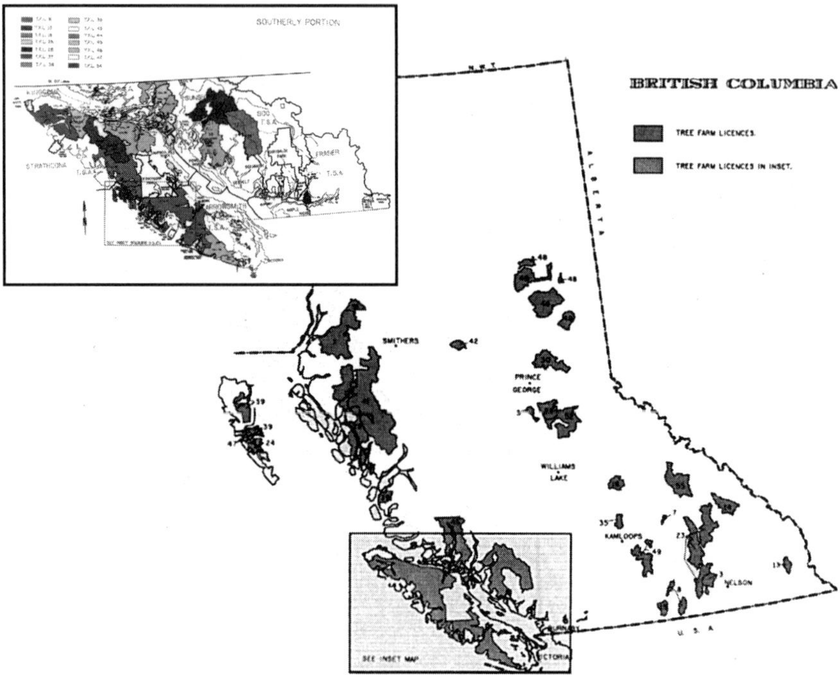

Source: *Forest, Range & Recreation Resource Analysis* 1994 (Victoria: Ministry of Forests).

In the Ministry of Forests' definition, mature forests are those in which the predominant species is older than the specified cutting age. The specified cutting age is bureaucratically established to meet the ministry's forest management objectives, which in turn are determined more by administrative considerations than by the state of the forest. In order to increase harvest levels, there has been a tendency to reduce the cutting ages of many timber types. For example, second-growth Douglas fir is deemed mature at eighty or eighty-five years though individual trees can grow and remain healthy for a thousand years or more.

Thus the designated "mature" forest area cannot be equated with old-growth forest. And "immature" forests are not always logged forest lands; some of them develop in the wake of wildfires or other natural disturbances. But there is an approximation between mature and unlogged land, and between immature and logged land. The area of

DISTRIBUTION OF TIMBER SUPPLY AREAS IN BC

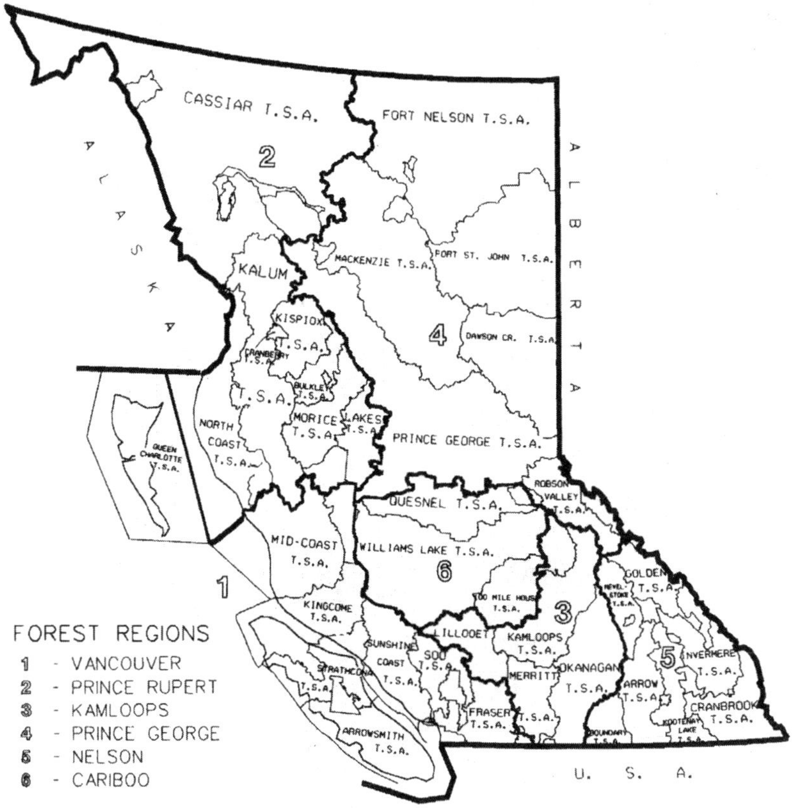

Source: *Forest, Range & Recreation Resource Analysis* 1994 (Victoria: Ministry of Forests).

"mature" second-growth forest in BC is not large: logging began in the 1860s, but substantial areas were not logged until some fifty years later.

Government inventory figures indicate that, contrary to the claims of some environmentalists and a growing public misperception, the last of BC old-growth forests are not about to disappear. At current harvest rates of about 200,000 hectares a year, it would take more than a century to log all the province's old-growth forests. Even in the heavily harvested Vancouver Region, there is more productive mature forest than immature. Much of the high-volume, easily accessible, high-value stands have been logged, and certain types of forests such as low-elevation coastal Douglas fir are scarce. Nor should the area of mature forest be confused with the area within which the forest industry can operate economically. At any given time, depending on the industry's technical capability, the cost of logging and the price of forest products, a portion of forest land is deemed inoperable. The Ministry of Forests, in making its allowable annual cut calculations, considers only timber volumes on the "Operable Land Base"—that portion of the productive forest "which is considered to contribute to and be available for long term timber supply." At the end of 1997, for instance, the provincial Operable Land Base was estimated at about 23 million hectares—19,692,573 hectares in TSAs and 3,212,546 in TFLs. These numbers should be accepted cautiously, because the ministry's use of "operable" is itself subject to a great deal of debate and controversy. At one time, for instance, a factor in reducing the operable area in the Mid-Coast TSA was the ministry's shortage of pickup trucks with which it could manage the area. However, Operable Land Base numbers do indicate that a substantial portion of the productive forest land outside parks and other protected areas is not likely to be used by the forest industry under existing technological and financial conditions.

Assuredly, the argument that we should stop logging all old-growth forests, or stop logging coastal old-growth forest, is debatable. Almost half of the productive forests in BC remain in their pre-industrial state. In some areas we are facing shortages of timber that is economical to harvest, but the forests are not in danger of disappearing.

Of greater long-term importance economically is the state of the new forest, which has replaced stands that have been logged off or burned over during the past century. There are more than 17 million hectares of new forest in BC, the bulk of which is available for future timber harvests. It is on these lands that our grandchildren's economic well-being rests. By far the largest portion of them is state-owned, and the job of managing them is directly or indirectly in the hands of the Ministry of Forests. How they are managed and to what ends is, therefore, a matter of great public interest. Management policy is not, as some bureaucrats and professional foresters claim, largely a technical matter.

Until a few years ago one of the great scandals of BC forestry was the amount of logged or burned-over land not satisfactorily restocked (NSR) with a new forest. As Table 1B (see Appendix) indicates, in 1995, 2.25 million hectares of government-owned NSR land was in Timber Supply Areas and Tree Farm Licences, most of it in TSAs. Another 1.33 million hectares were covered with non-commercial species. A zero-sum NSR is not realistic; an NSR area of 600,000 hectares more accurately reflects the delay between harvesting and regeneration. The difference represents a backlog of neglected land that has accumulated over the years, mostly because the Ministry of Forests did not reforest land for which it was directly responsible. Politicians have been unwilling to allocate the money needed for the job, and the provincial forest bureaucracy has functioned in a cumbersome manner.

Until 1987 the NSR area grew because the area harvested each year exceeded the area regenerated. That year, funds were provided by the federal–provincial Forest Resource Development Agreement (FRDA) and the backlog began to decline as more area was reforested than was logged. However, the current condition of the new forest is not clear and the volume of timber it is producing is a matter of dispute. Industry foresters have long been convinced growth rates are substantially higher than estimates prepared by the Ministry of Forests during AAC reviews.

In 1995, when a contentious review of harvest levels indicating a need for reductions in much of the province was under way, the MoF began a study of growth rates, the Old Growth Site Index project (OGSI). Within a year, preliminary results reinforced the industry position, suggesting growth rates were substantially higher than the ministry's previous estimates. Early results of this study caused much excitement in some quarters. Recent timber supply reductions may have been unnecessary, it was argued, and perhaps the harvest could even be increased in some areas. The final results of the study will be incorporated into timber supply analysis and AAC calculations when

OGSI is completed in 1998. Indeed, there are critics who view these findings as overly optimistic, and until there is a thorough examination of the project and its methodology, application of the findings in the AAC determination process will be contentious.

But even when the matter of supply is settled, the study does not provide complete information on the state of the young forest. We know very little about its health, its density and the quality of its timber. For example, for both new and old forests, the MoF publishes only gross estimates of the prevalence of insect attacks, which are calculated from aerial surveys, and diseases such as root rot typically occur in small pockets of an otherwise healthy forest. Because there is no land-based silvicultural workforce in these forests, there is no one on site to discover diseased stands or to treat them when they are found.

In the second-growth coastal and Interior forests, Douglas fir is often afflicted with Gilbertson root rot (*Phellinus weirii*), which is transmitted to new forests from infected fir stumps in logged or burned areas. As trees in a new forest grow larger and their roots interconnect, this disease can migrate from one tree to another at the rate of some 15 centimetres a year.

Parasitic dwarf mistletoe, which also spreads in slow-growing pockets, favours coastal hemlock. Small shoots grow on the branches and an extensive root system burrows down the trunk, weakening the tree. Each year the mistletoe produces a sticky seed, then in the summer it generates enough internal pressure to propel the seed 8 or 10 metres through the air. If the seed lands on a tree, it sticks and soon grows to infect its new host.

Treating these diseases is relatively simple and inexpensive if a silvicultural worker with the knowledge, skills and tools to identify and deal with them is at hand, perhaps working in the forest on some other task. But there are few such workers on most government-owned land in BC, so disease can spread unchecked until the damage is so severe the entire stand has to be treated or removed. The MoF acknowledges that the extent of forests affected by disease is not known, but according to its own estimates, root diseases are responsible for about 16 percent of timber losses each year, or about 2.2 million cubic metres of merchantable timber. These figures are only estimates, and only of merchantable timber. They do not incorporate infection rates in younger stands. And little of the infected wood is salvaged.

While no one knows how widespread the disease is or how fast it spreads, the vulnerability of most of the new forests is quite apparent. From the 1930s until just a few years ago, large clearcut logging sites were reforested with monocultural stands of only one tree species, often an inappropriate one. Some of these plantation forests, which replaced mixed-species stands, consist of hundreds or thousands of hectares of coastal Douglas fir or Interior spruce, which were the most commercially valuable species at the time they were planted. During the 1970s and 1980s, species diversity was narrowed even further when many of the younger stands were spaced in an attempt to accelerate growth rates. At that time, workers were instructed by the MoF to remove all but the most desired commercial species.

A 300-hectare pure stand of Douglas fir is far more susceptible to root rot than a stand that also contains, for example, cedar and hemlock trees, which are not affected by the disease and will not pass it along to vulnerable species. Red alder, not considered commercially valuable, was always removed even though it actively inhibits the spread of root rot. We now know that probably every monocultural forest in the province is susceptible to at least one pest or disease that can devastate that forest sometime during its growth cycle. No one knows to what extent the new forest is already under attack.

Forest diversity involves much more than number of tree species: any particular living forest community comprises a huge variety of animals and other plants. Given the fact that most of BC's new forests germinated in the wake of clearcut logging, with or without human help, and given our decades-long practice of growing monocultural forests, a wider issue of biodiversity arises.

As many studies show, there are good economic reasons for maintaining forest biodiversity in old and new forests. In their natural state, trees stay healthy in an environment in which many life forms are interdependent. Some species can be cultivated in isolation, but only at enormous cost and effort or on low-rotation plantations in highly accessible areas, such as on former cotton fields in the southern US. Growing timber in relative isolation from other forest flora and fauna, over the long growing period that BC species require, is a dubious proposition.

The biodiversity of BC forests is affected in two major ways by the manner in which the forest industry has operated in the past and to a great extent still operates. First, clearcutting large areas of forest has traumatized some resident plants and animals, which can be destroyed outright or forced to migrate. Second, we lack understanding about and have made few attempts in repopulating the new forest, within its planned rotation, with the various species that formerly lived on the site.

Although logging regulations that restrict the size of clearcuts were introduced in the 1990s, this is at best only a short-term, partial solution. If we cannot manage new forests that are as biologically diverse as the ones they replace, we will have to leave very large areas of refuge for the displaced species. This will severely curtail the size of the commercial forest land base. The forest industry is faced with the choice of either learning how to re-create biologically diverse new forests, or limiting its operations to a smaller portion of the forest land base.

Most of the new forests we have acquired so far were not created with biodiversity as one of the objectives. In some cases provisions were made for game species such as deer, elk and caribou, or other fish and wildlife populations, and recently measures have been adopted to protect the habitat of certain threatened species such as the northern spotted owl. However, the principal objective has usually been to replace the timber crop. It has been assumed or hoped that other forest creatures would wander back in and make themselves at home once the new seedlings had grown to an appropriate height—when those creatures were thought of at all. When an old-time industrial forester talks with pride about the healthy forests that have grown back in logged-off areas, he is probably referring to the trees. He may not know—or may underestimate—the importance of the woodpeckers that used to create cavities in old snags, which are used by other birds, which in turn feed on the bugs that eat the trees—the bugs that can now multiply unchecked and damage or destroy the forest. That is only one example of the interdependent relationships that maintain the health of a forest. Biodiversity is not just a matter of aesthetics: a more biologically diverse forest is a healthier forest, and a healthier forest is more productive ecologically and economically. Even if a new forest is well stocked with young, healthy trees of the most sought-after commercial species, many plants and animals may be missing, and others may survive only in drastically reduced numbers.

It need not be this way. New forests do not have to be so narrowly defined or parsimoniously designed; it is not necessary or advisable to crowd every growing space with tree species that fetch the highest prices on the current log market. In recent years, foresters have begun to recognize this. But they are still planting for the marketplace instead of for a healthy biodiversity. To achieve that goal we are going to have to acquire more knowledge about forests and biodiversity, gain more experience in applying that knowledge creatively, and provide more people with opportunities to sustain themselves while they work to sustain healthy forests.

An alternative to this kind of forestry is high-yield plantation forestry. Apart from a few relatively young hybrid poplar plantations, there were almost none of these plantations in BC by the late 1990s. Most of them are in the southern hemisphere or on reforested agricultural lands in the southern US, and on the Iberian peninsula. Whether they can be considered forests is a debatable question, since their primary purpose is to provide commercial timber: the management of plantations owes more to agriculture than to forestry so they should probably be evaluated by agricultural rather than silvicultural standards. One of the major virtues of timber plantations is that by efficiently providing high yields of fibre they reduce the need to harvest other types of forests. In a climate of controversy and pressure from all sides, much interest has been expressed in such high-intensity forest management zones.

Hectare for hectare, many of the new forests will yield a smaller volume of timber in 80- to 100-year rotations than the old-growth forests they replaced. Some of the old-growth stands that were "decadent" (no longer producing new timber) have been replaced by new stands that will produce higher volumes. Overall, though, we can expect to harvest less timber from each hectare of new 80- to 100-year-old forest than we harvested from the preceding forest. This effect is called "falldown." Some argue that we should head off the "falldown" effect by reducing harvest levels today to avoid timber shortages in the future. Others maintain such a policy only inflicts certain economic curtailment today instead of uncertain economic curtailment in the future. It is also argued that at some sites old-growth forests are so decadent that removing them and replacing them with young, healthy stands will increase timber supplies. While widely debated, the phenomenon of "falldown" has not been a significant factor in establishing AAC levels in BC.

Another question regarding new forests is wood quality. Old-growth forests are often dense stands where lower branches die at an early age and fall off, a form of self-pruning. Trees harvested from these stands during the first century of logging yielded a high proportion of clear, straight timber. The wood from the younger trees in the less dense new forests is often coarser grained, with less strength and more knots. The timber manufacturing sector has already undergone restructuring in response to this change. Some companies have adopted a strategy of developing "engineered wood products"—beams, panels and other items manufactured from low-grade logs—in order to add value to the available timber supply. Another way to add value would be to grow higher quality timber, but so far this approach has had little attention.

As timber supplies are reduced, we hear more and more about "value-added" manufacturing. In an example of this approach, instead two-by-four studs being sold into the export market for $250 a thousand board feet, the same volume of wood is used to make ten pieces of furniture that sell for $250 each. To sell wood in the form of products, rather than commodities, appears to make a lot of sense. But unfortunately, opportunities for adding value to timber from second-growth forests are severely limited by wood quality. On the lower coast, for example, there are thousands of square kilometres of second-growth Douglas fir forests. When these trees grow, particularly in well-spaced plantations, the branches gradually die off from the ground up. It can take thirty years for a dead fir branch to rot and fall off, after which the tree grows over the stub and begins to produce clear wood. Until that time, the wood that has grown around and over the dead branches is filled with loose knots, which fall out and leave holes when lumber or veneer is cut from it. Loose-knot fir can be used for making lower grades of lumber, pulp or veneer for sheathing-grade plywood. It cannot be used to make higher-priced furniture, moulding stock or marine-grade plywood. Therefore it is quite difficult to "add value" to this type of wood.

In 1995, in Vancouver, a sheet of five-eighths-inch sheathing-grade Douglas fir plywood containing 20 board feet of wood retailed for $23. A sheet of five-eighths-inch marine-grade plywood, made from clear Douglas fir, sold for $80. The difference in value of $57 a sheet, or $2,850 per thousand board feet, is the value we have lost by failing to manage new Douglas fir forests as well as we might have over the past fifty years. We have not had people working in the forests, clearing out the poorly formed, diseased and damaged trees and tending those selected to grow, pruning away branches and in other ways producing higher quality wood. It was an opportunity lost in the woods: the $2,850 cannot be made up in the mill.

In the last two decades of the twentieth century, a large amount of taxpayers' money was spent on specific silvicultural work in the young forests. Much more effort has gone into growing more timber than in growing better timber. At great cost, many thousands of hectares have been juvenile-spaced, and others have been fertilized. With luck, this work might increase growth by 15 or 20 percent. A 20-percent volume increase in loose-knot timber worth $250 a thousand board feet is not much of a return on investment compared to increasing the value of that thousand board feet by $2,850.

The real failure of BC forest policies, as revealed in the state of new forests, is not so much one of neglect as one of ill-conceived goals and forgone opportunity. Over the past fifty years the forest industry has not simply "cut and run," as many industry critics claim. Restocking has not always kept pace with harvesting, and we have much to learn about sustaining forests rather than mere tree farms. But generally where logging has taken place, or where fire has destroyed a forest, restocking has ensured new growth of a commercial timber species. Where we have failed miserably is in our inability to realize the economic opportunities available in the new forests of BC. That failure is a consequence of our policies, as shown in the history of their development.

2 A Century of Forest Policy

Making forest policy in British Columbia is a strange and often Byzantine process, rife with smokescreens and hidden agendas. It is always an intensely political process, underlined by the fact that vast fortunes are at stake and the lives of hundreds of thousands of citizens are directly affected by forest policy decisions.

For most of the province's history, forest policy was made through the periodic establishment of public inquiries, which solicited proposals from interested parties. The commissioners in charge of an inquiry then produced a report containing recommendations for dealing with current pressing forest policy issues. The government of the day picked and chose from among these suggestions, and proceeded more or less as it saw fit. The process was far from perfect, but it may have been more satisfactory than the method that came to replace it.

Between 1909 and 2000, five inquiries into the BC forest sector were held. The purpose of the inquiries was to engender a democratic, continuous, rational decision-making process and a progressive movement from unimpeded forest liquidation to enlightened forest management. In fact, BC forest policies have not been particularly enlightened, and the way they were arrived at has hardly been rational. They have, however, been a faithful reflection of BC politics.

The first inquiry into forest policy, called a Royal Commission of Inquiry, was appointed by the government of Premier Richard McBride in 1909. Fred Fulton, the province's chief commissioner of lands, sat as chairman. At that time there was no comprehensive body of forest legislation on the books in BC, no rules for determining how much could be harvested, how it could be harvested or what was to be done with forest land after harvesting. There were only a brief set of ordinances and a few sections in the Land Act defining how government-owned timber could be acquired by private interests. There was, however, a great deal of interest in forests, particularly on the part of a newly interested player—the provincial government.

On the coast, the Anderson sawmill in Alberni Inlet had come and gone almost fifty years earlier. The Moodyville and Hastings sawmills were operating successfully in Burrard Inlet, selling modest volumes of lumber into an export market operated through US brokers. The Victoria Lumber & Manufacturing mill at Chemainus was in the same business on a larger scale, and a few smaller mills, most of them on the Lower Mainland and Vancouver Island coasts, served local markets. In the Interior, a few small sawmills established during the gold rush days of the 1860s still operated in the Cariboo and east Kootenays. Several larger mills built along the mainline railways cut lumber for prairie and eastern rail markets. The forest industry had not been the main driver of the provincial economy and had contributed relatively little to government coffers.

That situation changed early in the twentieth century, when a timber boom erupted in BC almost overnight, primarily because of a series of events that took place in the US. For close to a century the North American timber industry had crept west, devouring the forests of eastern Canada and the US before leap-frogging across the American plains into the Pacific Northwest. In the late 1800s a conservation movement was born in the US, and the US government placed large areas of western American forests into reserves—not, as it is often claimed today, to protect them from loggers, but for future use by the forest industry in its role as provider of America's building materials. Suddenly there was a perceived shortage of timber in the US. There were still vast areas of untouched forest land that had been purchased from the railway companies. But speculators, who moved ahead of the forest industry, buying and holding forest land for future sales, were experiencing a shortage.

In BC, this sort of speculative activity was limited. About 3 million hectares had been granted to railway companies, most of it comprising the Esquimalt & Nanaimo (E&N) Railway lands on eastern Vancouver Island and the Canadian Pacific Railway lands in the Kootenays. A 4-million-hectare parcel extending 32 kilometres on each side of the CPR mainline across BC was ceded to the federal government when the province joined Canada in 1871. The result was that government still owned more than 95 percent of the provincial land base, and land speculation was not a significant part of the BC economy.

In colonial times, the policy in BC had been simply to give or sell forest land to investors wanting to build sawmills. This is how the Alberni and Burrard Inlet mills were begun by British financiers. The Chemainus mill was built by Americans and, typically, they chose to secure their timber by purchasing land from the Dunsmuir family, which controlled the E&N properties. As well, in 1865, a colonial land ordinance had been adopted allowing the Crown to grant timber harvesting leases on public land, while retaining public ownership of the land. Then, in 1896, the BC government passed legislation prohibiting the sale of public forest land. Forest land was defined as land west of the Coast Mountains carrying 8,000 board feet of timber per acre, or land east of the mountains carrying 5,000 board feet.

Since that time, defenders of government ownership of forest land have portrayed this legislation as an enlightened decision by public officials to ensure the proper care of the province's forests by keeping them in the public domain. There is little evidence that such care has been taken, yet it became one of the assumptions upon which a century of forest policy was based.

Until late in the nineteenth century there was almost no concern in BC, or elsewhere in North America, for the perpetuation of forests.

Forests were seen as an impediment to agriculture. Huge areas and enormous volumes of timber in the Fraser Valley, the Kootenays, Vancouver Island and elsewhere were cut and burned to create crop and grazing land. The common view across the continent was that a transient forest industry could be supplied from these land-clearing operations, after which people would get down to the business of farming. As H.V. Nelles explains in his history of Ontario resource politics:

> The nineteenth-century community did not look upon the forest as a living organism, perpetually growing, dying and regenerating itself. Rather the forest was perceived as a static entity, a unique historical event. People believed that once it had been cut over or burned, the forest could never renew itself on its own. Instead it would naturally deteriorate in stages of inferior weed and second growth. In summary, the nineteenth-century image of the forest was that of an enormous non-renewable resource, not unlike a giant mineral deposit, which was permanent simply by virtue of its size and could be exploited only once and then passed on to the farmers.[1]

BC's policy of state ownership was adopted from eastern Canada, particularly Ontario, where it had evolved in the early nineteenth century as a means of protecting strategically important timber supplies for Royal Navy shipbuilding. (Significantly, this was one of the basic principles the Americans discarded when they cast off the colonial yoke in 1776.) Colonial military authorities in Canada contracted private companies to harvest the timber they needed from their reserves. Then, as domestic and export markets for timber developed, and military needs declined, control of timber reserves passed into the hands of Canada's pre-Confederation governing bodies and, eventually, to provincial governments. A system of leasing the right to cut timber on the reserves evolved. At this point timber companies had no interest in owning forest land. In fact, the Canadian system suited them well: they did not have to pay for the timber until after it was logged, when they paid stumpage or royalty based on volume. Therefore they did not have to tie up capital in purchasing timber, as did the Americans, giving Ontario lumbermen a built-in competitive advantage. Here was born the Canadian system of government support for the forest products industry. A century later, the fundamentals of this system would still be a major bone of contention between the US and Canadian forest industries.

During the first half of the nineteenth century, Ontario farmers and lumbermen enjoyed a mutually supportive relationship. The farmers worked in logging camps during the winter, and in the spring returned to working the land that had been given to them by the government after it was logged. The government obtained most of its revenue from timber sales and could tax the farmlands at modest rates. This system worked to everyone's benefit until mid-century, when settlers began following lumbermen into the heavily forested land of northern Ontario—land of marginal agricultural potential. The lumbermen objected because of the fire hazard posed by settlers and squatters, and began agitating for a permanent commercial forest land base. A separate movement, which supported the sale of public forest land, arose at this time. It was suppressed by the provincial government with the strong backing of the established forest industry, which feared the consequences of having to purchase timber. As Nelles commented:

> The device of crown ownership and licensed rental gave the lumbermen access to the forest, returned a welcome revenue, and under ideal circumstances generated cut-over lands that could then be recovered from the lumbermen and passed on free to the homesteaders. The visible temporary tenure of the lumberman, expressed in his annual licence and ground rent charges, allowed the state to open up desirable lands when pressed to do so. Because it served the political and economic interests so well, crown ownership continued to serve as the basis of Ontario's forest law on into the twentieth century.[2]

Until the turn of the century, those who framed laws relating to forest land ownership in Ontario showed no concern about conserving forests, or providing for their proper management. The objective was to liquidate them as profitably as possible and turn them over to farmers. The motivation in BC in 1896 was essentially the same.

In 1905, hungry for revenue, the McBride government took advantage of the booming American demand for timber. The government rejected proposals to sell forest land and devised a new form of timber licence that was renewable and saleable and thus very appealing to speculators. It was attractive to legislators and taxpayers because it contained a provision to revise timber royalties annually, allowing government to capture a portion of the speculative profits earned from rising timber prices.

The scheme was a roaring success. Within less than three years, 15,000 licences had been issued, covering almost 4 million hectares of the province's best timber stands. The government was elated and alarmed. Enthusiasm for the new licences foretold a prosperous future for the BC forest industry and for the government. Then a new element entered the picture. The conservation movement was spreading from the US to Canada, and its proponents here were raising profound questions about the treatment of Canadian forests. Conservationism was a scientific movement based on the premise that forests were a renewable resource. The movement had originated in classical European forestry and was first introduced to the North American public through forestry schools at Yale and Cornell universities.

In 1906, excited by conservationist ideas, Prime Minister Wilfrid Laurier convened the first Canadian forestry convention in Ottawa, attended by 267 people from across the country. Most of the major forestry figures from the US attended, as did the handful of Canadians who had obtained forestry degrees from American schools. Basic concepts of conservationist scripture were well aired at the gathering. Speakers made it abundantly clear where they differed from preservationists, who existed in small numbers even then. Conservationism was about saving forests to use, about perpetuating them through a permanent commercial forest industry. Its primary concerns were efficiency and economy, not ecology, as Gifford Pinchot, the leading US forester of the day, emphasized: "We must put every bit of land to its best use, no matter what that may be—put it to the use that will make it contribute most to the general welfare...Forestry with us is a business proposition. We do not love the trees any the less because we do not talk about our love for them...use is the end of forest preservation, and the highest use."[3]

Bernhard Fernow, the German-born and -trained godfather of North American forestry, told delegates: "Forests grow to be used. Beware of the sentimentalists who would try to make you believe differently."[4]

A leading figure in the conservationist movement was Henri Joly de Lotbinière, a former premier of Quebec whose family owned a large forest estate on the St. Lawrence River. Joly de Lotbinière had served as president of the leading conservationist organization, the American Forestry Association, and in Laurier's federal cabinet, where he exerted substantial influence on early Canadian forest policy. From 1900 to 1906, as lieutenant governor of BC, he played a significant role in the development of the McBride government's forest policies.

A key element of the conservationist credo was state ownership of the resource, which led to the designation of national forests in the US at this time. As Nelles described this ideology, "Only the state possessed the necessary resources and jurisdiction to implement a sound forestry program. The task exceeded both the capacity and often the will of private owners who were, more often than not, concerned primarily with immediate returns. Moreover, the effects of combined individual neglect or short-sightedness redounded upon the entire community in the form of floods, droughts, exhausted soil and scarcity. Therefore, it was the duty of the state to protect the public interest against the rapacious instincts and ignorance of its citizens."[5]

The theory was that a technocratic elite of scientifically trained professional foresters could best protect the forests from wasteful logging, greedy politicians and the excesses of nature—fires, insects and disease. One of Canada's first professional foresters, Judson Clark, told delegates to the Ottawa conference: "Nature's method of waiting an age for the trees to disappear after they had passed their prime was wasteful alike in time and material. The forester with his axe saved the material and time."[6]

Not long after the convention, Clark resigned his position with the Ontario government, disgusted with its failure to adopt the kind of management policies he propounded. He moved to Vancouver, where he set up the province's first forest consulting firm and began staking McBride licences. He quickly established himself as the leading conservationist in BC, and his word carried considerable weight in the growing public debate provoked by the timber staking frenzy. At the end of 1907, the government stopped issuing new licences and, after discussing forest issues for a year, it announced the appointment of the Fulton commission to study the situation and make policy recommendations.

The real authority behind the commission was its secretary, Martin Allerdale Grainger. An itinerant Australian mathematician with a degree from Cambridge, Grainger had come to North America and headed west, stopping at the more renowned centres of learning on his way. There he encountered the conservationists, whose ideas he found convincing. He went on to the west coast and eventually ended up in Vancouver, where like many others he found work staking timber licences. He spent a year travelling by boat along the BC coast. Soon after, in pursuit of his wife-to-be, he found himself back in London, broke. To pay his passage back to Vancouver, he dashed off a novel, *Woodsmen of the West*, based on the handlogging community then thriving around Minstrel Island. Having demonstrated his facility with a pen, Grainger was given the job of commission secretary. The report he wrote was based on conservationist principles, and drew heavily on his experiences as a timber cruiser.

The Fulton Commission made twenty-one recommendations, most of which confirmed government policies already in place. Having determined that the McBride leases were sufficient to supply industry for the foreseeable future, the commission recommended maintaining the suspension on issuing leases and, where special circumstances warranted, make timber available at public auction in the form of Timber Sales. The last recommendation was adopted and eventually became the primary means of disposing of Crown timber. The report also called for the creation of a provincial forest service and advised the government to place timber revenues in a fund to finance forest renewal.

The province's first Forest Act, passed in 1912, incorporated most of the recommendations and encapsulated the basic notion of conservation—to renew forests in perpetuity. It also created a professional forest service charged with protecting the forest, collecting revenues and other functions. Further timber disposals were limited to small sales by public auction.

On the advice of Clark and Fernow, a young forester employed by the federal government, H.R. MacMillan, was hired as the province's first chief forester. One of his first acts was to broaden the forest service's role to include the rapid development of a provincial forest industry. It was the first but certainly not the last time the province's chief forest bureaucrat undertook a major policy initiative on his own authority.

In his first annual report, MacMillan described the rationale for linking the forest service closely with industry.

> The annual growth of the forests of British Columbia is even now, before they are either adequately protected from fire or from waste, certainly not less than five times the present annual lumber cut...What is not cut is wasted in the end. It is not merely advisable to encourage the growth of our lumber industry until it equals the production of our forests—it is our clear duty to do so, in order that timber which otherwise will soon rot on the ground may furnish the basis for industry, for reasonable profits to operator and Government, for home-building, and, in the last analysis, for the growth of British Columbia.[7]

It was an argument straight from the heart of the conservationist creed. MacMillan believed that without a large forest industry pumping money into the provincial treasury, the BC Forest Service would never receive the budget allocations required to care for the province's forests properly. His other significant achievement was to convince government that the hiring of Forest Service staff should be free of political patronage. He wanted a professional body of scientifically skilled managers, not a collection of political hacks.

MacMillan had barely begun implementing his plans when World War I intervened, and the fledgling forest industry collapsed when US lumber brokers stopped shipping BC lumber abroad. MacMillan was seconded to embark on a global trade mission to sell BC lumber, an indication of the prevailing priorities. Shortly after his return, he resigned from the Forest Service and was replaced by Grainger.

The war ended all thought of forest management in BC. The Forest Service had been established and the mechanisms for developing a forest industry were in place, but conservationist momentum died on the battlefields of Europe. A large number of Forest Service staff had enlisted, and many never returned. The forests bureaucracy that emerged after the war focussed its attention on forest protection, revenue collection and its own advancement. Reforestation was left to chance and no attention was paid to the new forests that managed to emerge on their own from the ashes of the old. Many of these young stands consisted of diseased or genetically inferior trees or were dominated by brush and non-commercial species. They were devastated by fire, some repeatedly, when slash fires or wildfires were started by logging operations and spread through them.

The 1920s was a decade of unprecedented industrial expansion in BC, with the forest industry leading the way. Development slowed during the early years of the Depression, but soon picked up again. Most timber harvesting took place on private and leased lands, particularly those held under the McBride licences. As logging activity picked up and a large number of smaller independent sawmills entered the booming export business, selling their lumber through the export company MacMillan established after leaving government service, more and more small, short-term timber auctions, called Timber Sales, were held. The only limit on the amount logged each year was that imposed by the marketplace. Coastal log production climbed from 1 billion board feet in 1916 to 2.8 billion in 1929 and 3.3 billion in 1940. These figures do not include the huge volumes of timber left to rot on the ground. Only the best logs were scaled for payment to the government and used in the mills.

This misuse of the forests did not go entirely unnoticed. MacMillan commented on it several times, observing that nowhere in Canada was anyone practising forest management. He also noted that provincial forest services, including the one he had founded in BC, had become more concerned with their own perpetuation than with the forests. In 1927 he wrote to a friend who was teaching forestry at the University of Toronto: "The public is still in the same mental condition, that there should be some forestry practised somewhere, preferably some place else. The public, so far as I have seen, has very little understanding of how the Government, to whom the duty of forestry is assigned, should practise forestry on public lands. The general assumption is that the practice of forestry by the Government on public lands is never going to interfere with the profit or comfort of the person who is thinking about it."[8]

It was a period in which policy was derived from practice, and in which practice centred on building an industrial economy during the headstrong years between two world wars. For much of this period very little thought was given to the future, particularly to a resource so apparently limitless as BC's forests. Almost twenty more years of unrestricted industrial expansion went by before anyone publicly expressed reservations about how the province's forests were being used.

In the late 1930s, BC's chief forester issued a warning about the consequences of allowing unregulated logging to continue without any forest management in place. He urged the government to implement the conservationist measures included in the Fulton Commission report of 1909.

It was in this same period that the first steps toward forest management were made by some of the companies that owned private forest land on Vancouver Island—notably MacMillan's company, which by now had acquired sawmills and timber holdings, and Bloedel, Stewart & Welch, the largest timber holder in the province. They introduced logging methods that would encourage successful natural regeneration, began planting seedlings on logged land and regarded the newly established young forests as their future timber supply. Early in the 1930s, the Forest Service started its first tree planting projects, as well.

These initiatives were overshadowed by the start of World War II, which, unlike World War I, led quickly to the mobilization of the BC forest industry. For the duration of the war, the forests were plundered with little thought of management for the future.

In 1942 the province's new chief forester, C.D. Orchard, took the first significant forest policy initiative since the Fulton Commission. He sent a confidential memo to Forest Minister A. Wells Gray, proposing a radical shift in policy with two central components: to regulate the harvest of timber from Crown lands, and to create a new form of forest tenure that would leave land ownership with the Crown.

Orchard's memo was sent in an extraordinary context. World War II was raging, with the Germans about to complete their conquest of Europe and the Japanese sweeping through the Pacific. A few days earlier, thousands of Canadian troops had died at Dieppe. The future of the country hung in the balance and the attention of everyone in the forest sector was focussed on winning the war. Orchard, who had been appointed chief forester eighteen months earlier when the then-chief, E.C. Manning, was killed in a plane crash, chose this moment to propose the most sweeping changes in provincial forest policy since the first Forest Act was passed in 1912. The consequences of his act have bedevilled BC ever since.

The idea of sustained yield proposed by Orchard was not new. It was the central tenet of conservationist doctrine at the turn of the century, and had been propounded by Fernow and his disciples for almost fifty years. Starting with this familiar concept—to ensure a steady flow of timber for the future—Orchard introduced mechanisms for regulating the annual timber harvest. His proposal was based on a form of "partnership" between industry and government. The Crown would retain forest land ownership and provide long-term leases to forest companies, which would manage the new forest established after the original one was logged.

The forest minister passed Orchard's memo on to Premier John Hart, who circulated 'it quietly among his friends in the forest industry. As Orchard later wrote, Hart "called me to his office, when he told me that the Government was 'sold' on my proposals, but that he couldn't hope to get such a radical change of policy through the legislature if it were introduced 'cold.' He, therefore, proposed to appoint a Royal Commission, he thought one man and hoped that it would be the Chief Justice, Hon. Gordon Sloan, to canvass the proposition, both for the value of his findings and as a measure of public education."[9]

The first public inquiry on forestry had been called in the wake of a timber boom and in response to British Columbians' growing interest in conservationist ideas. This one was called not to develop policy, but to sell the chief forester's administrative concepts to the public.

Sloan was duly appointed. He conducted the requisite public hearings and produced his report in December 1945. By and large he accepted Orchard's proposals, and in some ways he went far beyond them, particularly in regard to the importance of non-timber forest values. He advised that the primary objective of provincial forest policy should be "to so manage our forests that all our forest land is sustaining a perpetual yield of timber to the fullest extent of its productive capacity. When that is accomplished all benefits, direct and indirect, of a sustained-yield management policy will be realized; providing, of course, that the multiple purpose of our forests is recognized as an aim as important as balancing cut and [growth]."[10]

To facilitate a yield regulation system, Sloan outlined a method for calculating the allowable annual cut but, significantly, did not suggest reductions in the current cut. He proposed a scientifically based system of cut controls designed to cultivate a provincial forest consisting of healthy, productive stands of trees gaining optimum volumes and values annually. To this end, he recommended that mature coastal forests be cut much more rapidly than in the past, then replaced with young, vigorous stands.

Anticipating government action by fifty years, Sloan also called for restrictions on logging methods used on private land, not to protect the environment, but to protect the future of the forest industry: "There comes a time when the public welfare must take precedence over private rights. To permit the owners of Crown-granted lands to log them off and leave them without taking any steps to secure the growth of a new crop is to jeopardize seriously the future development of our logging industry."[11]

Sloan's comments on tenure were somewhat inconclusive. He did not adopt Orchard's detailed concept of forest management tenures but underlined the need for private forest management and mentioned combining Crown and private lands into forest management units. He also described "private working circles," large or small areas of forest land that could supply particular mills in perpetuity, while "public working circles" could be managed to supply timber to the open log market. He also called for another Royal Commission on forestry to be conducted ten years hence.

Sloan's recommendations for sustained yield forestry were favourably received by the public, industry and government, and a new Forest Act based on them was drafted in 1948. A system of public and private working circles was established, the first called Public Sustained Yield Units (PSYU), and the second, Forest Management Licences (FMLs). Orchard commented later that although Sloan had not recommended his suggestion of Forest Management Licences, he had not specifically rejected it either, so the provision appeared in a draft copy of the new legislation. Unlike earlier tenure provisions, which had provided the right to harvest a specified volume of timber, the new licences consisted of leases on a specific area of forest for a perpetual term, including the right to harvest the timber in that area and responsibility for the forests in it.

The industry was caught off guard by the inclusion of these unexpected tenure provisions and was unable to organize much opposition in the two-week period the government allowed for discussion. The new laws were passed with little modification, including the provision for FMLs. It turned out to be the most controversial piece of forest legislation ever passed in the province.

Several senior members of the Forest Service resigned, believing the tenure plan was wrong and Orchard was acting autocratically. A group of private sector foresters and industry leaders opposed the idea in favour of a proposal to transfer logged-off lands into private ownership and requiring they be managed for future timber crops. MacMillan was at the centre of this group, but he was the only big operator to support the small independent loggers in their opposition to the FML plan. They contended the measure would lead to a concentration of control over cutting rights and, because it would not provide the required incentives for full-scale forest management, would also lead to future timber supply shortages, mill closures and damage to local economies.

Orchard had intended to award modest-sized FMLs to several hundred of the more forward-thinking logging companies, which could be persuaded to begin the task of growing future timber supplies. Before the ink on the new legislation was dry, however, the government began awarding very large areas of unlogged forest land to large US corporations with no previous presence in the province. The first FML was granted in Forest Minister E.T. Kenney's riding, giving the Celanese Corporation of New York a perpetual lease on 809,400 hectares of forest. This licence would plague governments for decades, eventually including a $200 million bailout by provincial taxpayers in late 1997. The second FML was granted to another US company on lands previously reserved for independent market loggers. Within months the industry was in an uproar.

By the time the W.A.C. Bennett Social Credit government was elected in 1952, the public had begun to realize that a few large companies were gaining control over large areas of land and huge volumes of mature timber through the new management licences. In response to a growing clamour, a second Royal Commission was set up in 1955, again chaired by Chief Justice Sloan. Although it was established with a broad mandate, the focus of the commission's deliberations was the issue of Forest Management Licences.[12]

During the commission, evidence was presented suggesting Social Credit Forest Minister Robert Sommers was accepting bribes in exchange for awarding FMLs. Sloan refused to accept the evidence, but this revelation and events stemming from it competed for attention in the press, and the drama continued to unfold at the commission hearings.

MacMillan appeared before Sloan several times and spoke strongly against granting any more FMLs in the most important areas of the coastal forest. He was challenged vigorously by the heads of major forest companies, who had filed applications for FMLs. In support of small independent operators, MacMillan predicted the new tenure system would lead to monopoly control of the province's forest lands by companies that in the end would fail to manage the forests properly. The irony of MacMillan—the province's first chief forester, who had become owner of the province's largest forest company since his departure from government—attacking head-on the centrepiece of the current chief forester's radical new forest policies, was not lost on the public.

Sloan's second report vindicated the FML system, without mentioning any of the controversy surrounding allocation of licences or responding to any of MacMillan's well-founded arguments against extension of the system. In one curious section, Sloan quoted verbatim his questions and Orchard's responses on the issue of ownership of forest land:

> **Sloan:** Is it a corollary of [Orchard's admission that private forest management was better than that provided by the Forest Service], then, that the Crown should gradually release more of its Crown-owned lands, say, by form of outright sale and Crown grants to companies who are prepared to take over a sustained-yield management?

> **Orchard:** Once again a purely personal opinion, I think perhaps it would be in the public interest if 50 percent of our lands belonged to private owners.
>
> **Sloan:** It would appear from what you say that the policy that has been enunciated here since about 1909 has discharged its function for which it was designed and we may be approaching a new concept in forest ownership and pattern in British Columbia.
>
> **Orchard:** I'm inclined to think so. I'm inclined to think that some of those statutes and policies that were adopted fifty years ago were wise beyond their day; they were extremely wise, valuable. I think that at our present stage of development one-half of our lands in private ownership would be a wise step. I don't think we can do it. I don't think we ever will. We've got ourselves involved, we've sold the idea of Government ownership so thoroughly, we have such a strong minority element of socialism, I don't think the people will ever let us sell the land. But those things aside, if we were just looking for the greatest benefit for the most people over the longest period of time, I think that we should dispose of one-half of our land in private ownership, and the rest of it, if it were possible, into forest management licences.[13]

It was rather an astonishing admission from the person who for a decade was the leading opponent of private forest land ownership in BC, as well as being the architect of an alternative land tenure system. But his observations were lost in the rapidly evolving course of other events.

Discussion of Sloan's report and the issues it dealt with was overshadowed a few months later when Forest Minister Robert Sommers went to trial on charges of accepting bribes for issuing FMLs. The chief financial officer of BC Forest Products Ltd. testified his company had paid money to intermediaries for an FML; shortly thereafter,

BCFP president Hector Munro died suddenly and was widely believed to have committed suicide. This dramatic event effectively ended the controversy over Forest Management Licences and the industry focussed its attention on making the best of the new system. The FML was renamed the Tree Farm Licence (TFL) and only a few more were issued. The province settled down to another two decades of steady growth of the forest industry under the terms of the 1948 Forest Act.

During this period, under the new forest minister, Ray Williston, the Social Credit government instituted a number of major policies without consulting the public or, in many cases, the industry. Williston's preferred method of making new policy was through regulations, which, unlike legislation, did not require acts of law. For sixteen years he served as forest minister, overseeing the transformation of the industry—very little of it accomplished through the Legislature.

The term of an FML was reduced to twenty-one years by government decree. As competition for timber in the Public Sustained Yield Units (PSYUs) intensified and cutthroat bidding became common, the government granted "established" Timber Sale licensees preferential bidding rights under a quota system that operated with no legal basis. Within a few years the government simply stopped issuing FMLs. And it encouraged the development of an Interior pulp and paper industry, in part through the creative use of existing tenure laws and by developing new forms of tenure on an ad hoc basis. The period was also characterized by the concentration of cutting rights into the hands of an ever-declining number of ever-growing forest companies. The de facto policy that emerged from two decades of continuous administration, most of it under Williston, was to encourage the growth of a few large integrated companies at the expense of many small independent firms, and to eliminate competition for government-owned timber. A corollary of this policy was a reduction in revenues collected from the sale of government-owned timber. At the same time, there was a steady increase in the Allowable Annual Cut, allegedly based on better inventory data, but likely based more on increased market demand and mill capacity. None of these were official policies, having escaped public debate and proclamation, but they were actively pursued through the 1960s and 1970s.

During most of this period, BC was governed by the Social Credit government of W.A.C. Bennett. But in 1972, BC's first New Democratic Party government was elected. The NDP had promised during the campaign to expropriate the forest companies, but on entering office found this was impractical. Having no other forest policies on its agenda, the government appointed a Royal Commission. Peter Pearse, a forest economics professor in the University of BC faculty of forestry, was named commissioner in 1975. He found himself in an unusual position when, as he was about to prepare his report, the NDP was defeated in a general election. Having no political expectations to fulfill, he was free to express himself fairly independently. After holding extensive hearings throughout the province and receiving submissions from 205 individuals, companies and organizations, Pearse prepared a lengthy report providing an exhaustive overview of existing policies, a discussion of their strengths and weaknesses, and scores of recommended changes.

Having been specifically instructed to examine tenure policies, Pearse devoted much of his report to this complex issue. What he found in studying the industry was that since 1954 control over timber cutting rights had become highly concentrated in the hands of a few large companies. Twenty years earlier, at the beginning of the FML era, the largest ten companies had controlled 37 percent of a 26.6-million-cubic-metre annual provincial timber harvest; by 1975 the harvest had grown to 60 million cubic metres, of which the ten largest firms now controlled 59 percent. A similar concentration was taking place in the various mill sectors as well. "In my opinion," Pearse wrote, "the continuing consolidation of the industry, and especially the rights to Crown timber, into a handful of large corporations is a matter of urgent public concern."[14]

He described in detail how this concentration had come about. Most of the FMLs, now called Tree Farm Licences (TFLs), had been acquired by the larger integrated companies, which had also begun acquiring cutting rights in the PSYUs. Within these public working circles an informal quota system had evolved and had taken on the characteristics of a legally binding arrangement, although it had no standing in law. In fact, as Pearse observed, it contravened the Forest Act. Under this system, the Forest Service simply apportioned the allowable annual cut of each PSYU to the area's "established operators" and gave them preference in the bidding process. The system ended competition for timber and favoured the bigger companies, which soon prevailed in the struggle for PSYU timber.

This domination of the industry by fewer, larger companies occurred in different ways and for different reasons on the coast and in the Interior. At the beginning of this process, a substantial portion of the coastal industry consisted of small independent market loggers who, because of the ease of transporting logs along the coast, were able to supply numerous independent sawmills. Few of these loggers or lumbermen were enthusiastic about relinquishing their cutting rights and their independence. In the Interior the sawmill industry, which had previously consisted of thousands of small, portable bush mills, was consolidating in order to channel its waste wood into a new and growing network of pulp mills, and the independent operators there were more willing to sell their cutting rights. In the end the result was the same: a few large companies held sway over the industry in both regions.

Pearse recommended leaving the tenure system essentially the way he had found it, except for opening up opportunities for small independent firms. Although he did not attempt to turn back the clock on the distribution of tenure, he did propose some broad changes in the laws pertaining to tenure. He recommended phasing out several of the old forms of leases and licences and reissuing them in fewer forms, more consistent with current needs and practices.

One of his recommendations for achieving a better balance was for the government to provide Woodlot Licences—parcels of forest land up to 400 hectares—on which smaller firms could practise forest management. The concept was spelled out in a brief presented to him at a public hearing by Jim Collins, a veteran BC forester. It identified more than 4 million hectares of second-growth forest land in the PSYUs that could be parcelled out in economically viable, small-scale forest farming operations.

In a general discussion of the fundamentals of forest tenure, Pearse delved into the matter of land ownership. This issue had rarely surfaced during his public hearings, so after a lengthy review of the question, he concluded in his report that there was no public pressure to change existing ownership policies. Orchard's observation about the effectiveness of public-ownership propaganda had been accurate. Industry was content with the prevailing system because it did not have to tie up its financial resources in purchasing and holding timber. Pearse reasoned that public ownership enables the government to protect and enhance non-timber values, and it provides the government with a powerful means of influencing economic development, so he recommended no change in the general policy of retaining Crown title to forest land.

Overall, Pearse's recommendations focussed on two main policy areas: to adjust the balance between large and small firms in the industry, and to abandon the narrowly conceived yield regulation system and its objective of creating an idealized forest with an even distribution of age classes, the so-called "normal forest." Instead he recommended that the emphasis be on maintaining the health of BC forests, including non-timber forest values:

> It is no longer adequate to fix the goal of yield regulation as a more or less constant flow of timber volume over very long periods. Present circumstances call for a more flexible approach to yield regulation, with greater emphasis on protecting and enhancing the productivity of forest land and on the economic, social and environmental implications of harvesting. Forest management policy for the future should be directed toward two related objectives: protection and enhancement of the capacity of forests to produce their potential range of industrial and environmental values, and within that framework the regulation of harvesting to produce the

maximum long-term economic and social benefits from the timber resource.[15]

The new Social Credit government, elected in 1975 and led by Premier Bill Bennett, adopted most of Pearse's recommendations, in some cases with just enough of a twist to protect the privileged status of the large corporations. For instance, where Pearse recommended fifteen-year renewable terms for TFLs, the government set the term at twenty-five years. Mike Apsey, a senior staff member with the Council of Forest Industries (COFI), an organization of large forest companies, oversaw the writing of the revised 1978 Forest Act, and then was appointed deputy minister of forests to implement it.

Existing tenure arrangements were modified to entrench the status quo more firmly, although provisions were made for a Small Business Enterprise Program (SBEP) that gave small independent firms the exclusive right to bid on a fixed portion of timber each year. Provisions were also made for the Woodlot Licences Pearse had advocated. Yet the government was slow to implement these provisions. It took almost twenty years and two new governments to bring the SBEP timber allocations up to 15 percent of the provincial harvest, although the level originally promised had been 25 percent. And by 1997 less than half of the approximately one thousand Woodlot Licences promised in 1978 had been allocated.

The Bennett government used Pearse's recommendations on yield regulation to push allowable annual cuts even higher, to levels that were neither sustainable in the long run nor supported by the timber-growth productivity gains Pearse had said were possible and needed. Major expenditures on silviculture, particularly those activities that would justify higher cut levels, were approved. The Forest Service was instructed, in essence, not to employ its old system of yield regulation.

The government also adopted a series of measures introducing a policy of multiple use. The stated objective was to maximize the benefits obtained from forests by integrating the management of the various resources, as opposed to maximizing the use of a single resource.

It sounded fine and was consistent with Pearse's recommendations, but in practice it often turned out to be quite different than its proponents had imagined. Designating an area of forest as "wilderness," for example, could now be rejected on the grounds it was a single use, whereas clearcutting the area could be justified as one of many uses, including such activities as recreation and fish production. In a cartoon from that period, a pair of hikers stand in the middle of a vast clearcut with a forester, who tells them: "Okay, we're finished. Your turn now."

Although the Bennett government incorporated most of Pearse's recommendations in one form or another, some of the new forest policies that grew from the exercise were a perversion of the proposals, not so much in the way the legislation was written, but in the way it was applied.

An example was the practice of "sympathetic administration," a set of arbitrary decisions made during the economic recession of the early 1980s to give financial relief to the major forest corporations. The Forest Service was ordered to refrain from enforcing certain parts of the Forest Act, particularly those concerning timber utilization. For instance, lower grades of merchantable logs were left in the slash or burned in piles at roadside. These and other subterfuges, none of which were made public, circumvented the spirit as well as the letter of forest law.[16]

Under the second Social Credit government, the big companies and their control of timber rights continued to grow. Independent market loggers all but disappeared, and the surviving companies continued to consume each other. With the cost of silviculture deducted directly from stumpage, which made it an expense paid out of the public treasury, timber companies made no real investment in the new forest. They expanded by acquisition, rather than by investing in forest management activities that would increase the volume and value of the publicly owned forests.

As the allowable annual cut rose and logging kept pace, BC experienced a series of confrontations between loggers and people concerned about non-timber forest values. Forest companies found themselves engaged in a growing number of disputes over their right to log the timber government had awarded them. The Forest Service, though accustomed since 1912 to its role as facilitator of the forest industry, found itself in the unwelcome position of having to take into account the views of non-timber interests.

By 1982, British Columbians were not the only people becoming alarmed at the theory and practice of provincial forest policy. US lumber producers, hurt by the recession, found their domestic markets smothered by a flood of inexpensive Canadian lumber, particularly BC lumber. They blamed the stumpage appraisal policies of BC and other provinces, and launched a countervailing tariff action, the first such attempt in more than twenty years. When that failed, they launched a second one. This attempt promised to be successful and the government responded by making fundamental changes in provincial stumpage policies. Now, policy was being made in response to circumstances of the moment.

In 1987 Dave Parker, the forest minister in the year-old Social Credit government of Bill Vander Zalm, touched off a province-wide furor over tenure. Parker was Vander Zalm's second forest minister. The first had been Jack Kempf, a political maverick who publicly voiced his lack of sympathy for the big forest companies and announced his intention to restore genuine free enterprise to the industry. Kempf had also abandoned the unified Canadian stonewalling of the American countervail attempt, and had precipitated the final solution to the dispute, which eventually entailed a large increase in BC stumpage rates. Kempf was then forced to resign over alleged misuse of his office expenses, and Parker was appointed to replace him.

Parker was the first professional forester to occupy the minister's office. He had worked on the infamous TFL #1 in northwestern BC, arguably the most poorly managed licence in the province. He was an unabashed supporter of the major forest corporations, who welcomed him to office with open arms. After serving as minister for only a few months he announced his intention to convert most of the existing volume-based cutting rights in the Timber Supply Areas into area-based Tree Farm Licences. His announcement provoked an outcry from within and outside the industry, so he set out on a tour of the province to explain the merit of his proposal. Here was a uniquely new method of making forest policy—a ministerial announcement of intent and a drum-beating tour of the province.

At the half-dozen public meetings he staged in various parts of BC, Parker encountered solid and widespread opposition—not only to his proposal, but to his decision to make policy by ministerial decree. Ray Williston might have been able to get away with such actions, but the times were different and the new minister was no Ray Williston. Lacking the diplomatic skills his position required, Parker argued with opponents and took an uncompromising stance on the issue of tenure conversions. The crowds grew bigger and more hostile as he went from meeting to meeting, and participants took the opportunity to berate the minister on a variety of issues, including tenure. His tour ended at a raucous meeting in Parksville where the meeting hall was surrounded by several hundred people demanding, among other measures, a Royal Commission on forest tenure. Parker had repeatedly rejected suggestions and requests for an independent public inquiry on the subject. The people of the province were accustomed to full and open public discussions of major forest policy issues before change was made, and were not about to accept the pronouncements of a high-minded, heavy-handed forest minister without an argument.

A few weeks later, on June 29, 1989, Parker called a press conference and sullenly announced the formation of an altogether new type of advisory group, a thirteen-member Forest Resources Commission (FRC). It was to be a permanent body that would provide the forest minister with advice and recommendations on whatever issues were referred to it. This was an old idea that had first surfaced in the Fulton Commission report and had been raised again by Sloan. A few months after the FRC was established, Vander Zalm dumped his second forest minister and appointed a third, Claude Richmond, who shortly thereafter mandated the FRC to examine and report on a broad range of questions, including the Tree Farm Licence system and logging practices.

The FRC was a cumbersome entity, most of whose thirteen members were also engaged in other occupations. The day-to-day management of its inquiry was handled mostly by the second FRC chairman, Sandy Peel, a career bureaucrat who had served in various provincial government departments. He had learned about the forest industry while serving as the province's watchdog over semi-secret government payments to the Council of Forest Industries.

The Parker debacle stirred up intense public interest in forest issues, and this interest found focus in the FRC public hearings. Some 2,000 individuals and organizations made formal presentations, almost ten times the number that had appeared before Pearse, and thousands attended the public sessions. In April 1991, the FRC presented its report to the minister. It disappointed almost everyone.

The report was uncompromising in its description of the state of the forest sector. The commission found that control over harvesting rights had become even more concentrated since the Pearse report in 1976. By 1990 the ten largest companies controlled 69 percent of an annual cut that had grown to 63.8 million cubic metres. During the same period, the cut controlled by the largest company, MacMillan Bloedel, had increased from 13 percent of the total to 17 percent. Eighty-six percent of the provincial timber harvest was controlled by only twenty of the largest companies, which had controlled 74 percent fourteen years earlier.

The report indicated concern over the use to which the large companies were putting the timber they logged: "The vast majority of wood is tied up under tenure by companies that primarily produce low value commodity products such as dimension lumber and market pulp. People with new ideas for wood products that would create more value and perhaps employ more people in their manufacturing cannot pursue their ideas because they are denied access to timber."[17]

The commission also expressed alarm about the timber supply and annual harvest levels, but deferred judgement on this matter because of a lack of reliable forest inventory information, a situation it termed a "disgrace."

The FRC concluded that what was needed in BC's forests was a refocussing and restructuring of the forest industry based on a concept of "enhanced stewardship," or intensive forest management. This idea went beyond the old notion of timber management and involved integrating management for all values on the largest possible forest land base. To realize the plan, the FRC stated, volume-based tenures in the Timber Supply Areas would have to be terminated and replaced by new area-based tenures. Forest land should remain in public ownership and all be leased to the private sector for the purpose of forest management, including the growing of commercial timber crops. The large integrated companies would be permitted to control tenures providing no more than 50 percent of their timber requirements, the intent being to establish a degree of separation between the timber harvesting and manufacturing sectors.

To this point, the FRC recommendations reflected fairly faithfully the ideas and opinions of those who had made representations to it, apart from the large integrated forest companies. Then the report went off the rails. To implement its tenure reforms, it called for the creation of a Crown corporation, the Forest Resources Corporation (Forestco), which would be given title to all Crown forest land. This corporation would be empowered to raise money, using government-owned timber as security, and enter into twenty-five-year forest management contracts with the area-based tenure holders—established Tree Farm Licence holders, community groups, Native bands, woodlot operators and other undefined groups and individuals. The corporation would also manage some of the land itself.

Response to the Forestco proposal was almost entirely negative. Critics predicted that it would create more bureaucracy. The idea of placing the province's forests in the hands of a Crown corporation vested with the power to borrow money on the future value of the forests alarmed practically everyone with a stake in the forests—except for those who saw themselves as potential functionaries in the new agency and its surrounding bureaucracy.

The Social Credit government, which was facing a general election later in 1991, quietly ignored the report. BC voters brought in a New Democratic Party government, which left the report on the shelf, ignored the FRC and eventually abolished it. Thus ended the province's experiment with a "permanent" commission as a device for shaping forest policy.

By the time the NDP assumed office in 1991, the backlog of forest sector problems had grown to enormous proportions. The annual timber harvest was grossly overcommitted in some areas, and the prospect of drastic reductions loomed in several Timber Supply Areas. As loggers moved into more remote and sensitive areas in search of timber, land use conflicts increased. Public concern about the way the industry operated was expressed vociferously. A series of court cases involving Native rights brought to the foreground the long-simmering question of Native land claims, an unresolved issue of potential significance for forest tenures. And a beleaguered forest industry, battered by weak markets and employing 100,000 workers with a propensity to vote NDP, was demanding clear direction on where and under what terms it could operate.

The new government set about dealing with problems its predecessors had allowed to fester. The Ministry of Forests was instructed, by legislation, to conduct by the end of 1995 an exhaustive review of the timber supply in every Timber Supply Area and Tree Farm Licence in the province. With this information, the chief forester would set new allowable annual cuts for each. As the first reviews were completed, the industry went into a state of shock. The Mid-Coast TSA was facing a 34 percent reduction in AAC. The Sunshine Coast TSA, with several timber-dependent communities and a large forest industry workforce, was confronted with a 24 percent reduction. Predictions suggested a province-wide reduction in the range of 15 to 30 percent of the annual BC timber harvest, which in March 1993 stood at 71.3 million cubic metres. It did not take a mathematical genius to figure out that reductions of this scale throughout BC would cost between 15,000 and 30,000 direct jobs, and twice as many indirect ones. Forest workers across the province reacted with alarm, and the government began to get nervous.

The review process became complicated when Chief Forester John Cuthbert announced the first AAC revision for a Tree Farm Licence, a reduction at MacMillan Bloedel's TFL #44. The company took the issue to court and obtained a ruling that the chief forester lacked the authority to order a reduction in AAC. The government soon changed the law, giving him that authority.

The matter of establishing new allowable annual cuts did not involve any policy changes, just a straightforward implementation of long-established policies. The chief forester now had the authority (at least in TSAs) to determine annual harvest levels through a clearly defined process. The previous government had prevented him from doing his job; this one simply told him to proceed and passed a law giving him a completion date.

However, there was no clearly established policy for resolving conflicts over land use, and no effective mechanism for doing so. The planning processes in place had been derived from principles of multiple use and integrated resource management, and they were inadequate for the task. If anything, with the domination of these processes by bureaucrats in various government departments who were accustomed to working in isolation, they made the situation worse. Industry and non-timber interests alike grew increasingly dissatisfied with them.

As well, the government was committed to establishing new parks and wilderness areas in the province, to a total of 12 percent of the land area. This was a campaign promise, derived from a United Nations-supported objective, that Premier Mike Harcourt had made in order to secure votes from environmentalists and other proponents of park land. He was also responding to economic and political pressure from outside BC to preserve more forest.

In 1992 the government set up the Commission on Resources and the Environment (CORE) under commissioner Stephen Owen to develop regional land use plans that would take social, economic and environmental issues into account. CORE began by setting up forums for representatives of various interest groups and government agencies in four regions—Vancouver Island, Cariboo–Chilcotin, East Kootnay and West Kootenay–Boundary—to develop regional land use plans. To encourage participants to be diligent, Owen made it clear he would recommend a plan to government himself if the regional groups failed to agree on one of their own. By late 1994, all four groups had come up with plans. They were incorporated by Owen in his report to the government, which used them as a basis for a series of land use plans announced for each region. Implementation of the plans was another matter: some of them still had not been undertaken by 1997, by which time CORE had been disbanded.

CORE was based on a faith in the efficacy of the planning process as a means of resolving resource use conflicts. It assumed that if all those with an interest in the issue were brought to a table and convinced their deliberations carried weight with government, they could resolve their differences or come to a workable compromise. During the first half of the 1990s it appeared that in parts of BC, everyone with the remotest interest in forests was crowded around one of many tables attempting to come up with a plan.

As an ongoing forum for resolving land use conflicts, CORE was an obvious failure: after completing its initial chores it was terminated, and its recommendations to establish a resource planning bureaucracy apart from existing government ministries was ignored. On the other hand, for many people with concerns about forests, it provided a seat at the table where the planning process was conducted. At the very least, CORE brought representatives of the forest industry together with members of several communities who questioned or were opposed to industry's activities. For all concerned, a certain amount of de-demonization resulted and people with conflicting opinions got to know each other and exchange views. CORE also brought the planning process into the open, from behind the closed doors of the various ministerial bureaucracies that had, in earlier times, jealously guarded it as exclusive turf.

In 1992 the government also initiated the Protected Areas Strategy (PAS) in response to a growing public demand for increased wilderness, parks and protected forests. This program incorporated two earlier Ministry of Forests initiatives, the Old Growth Site Index project and Parks and Wilderness for the '90s, a joint endeavour with the Ministry of Environment, Land and Parks. These programs, as well as the PAS itself, involved massive consultation between the forest industry, other forest-user groups and the public. The following year, a bill to double the area of park land to 12 percent of the province was passed in the legislature. Shortly after the PAS was adopted, several large new parks were created at Tatshenshini, Khutzeymateen, Ts'yl-os, Nisga'a, Kitlope and Clayoquot Sound, most of them removing land from the commercial forest base and reducing allowable annual cuts in the Timber Supply Areas affected.

Well before the NDP government took office in 1991, the idea of some sort of code to govern logging methods was under discussion. The Ministry of Forests began drawing up a draft code, which at that point was seen only as a set of guidelines. In the same vein, the Forest Resources Commission recommended adoption of a forest practices code that would be enforceable by law. In the meantime, a European campaign to boycott BC forest products was gaining momentum, a situation that gave a common cause to the NDP government, the major companies exporting forest products and the forest sector unions. The government announced its intention to impose a forest practices code on the industry and turned over the task of framing a code to the Ministry of Forests. At this stage, most people in the industry were in favour of some sort of code that would define standards for the way they operated in the woods, if only to forestall the threatened European boycott.

In 1993 the NDP's second forest minister, Andrew Petter, and the minister of environment, Moe Sihota, released a discussion paper outlining in detail the government's plans for a code, backed by a staff of enforcers and financial penalties of up to $1 million per violation. The code itself would consist of a new set of laws, a body of regulations and a set of guidelines. It would be administered by a beefed-up Ministry of Forests staff and overseen by a new Forest Practices Board and an appeal commission. In its emergent form, the code was seen by most people in the forest industry as a bureaucratic monstrosity, only marginally better than a European boycott.

The adoption of the Forest Practices Code was an instance of policy-making under fire. Its primary objective, for government, industry and the unions, was to convince European customers that BC's forests were not being pillaged to provide them with two-by-fours and toilet paper. To be successful, the code had to appear to be tough and uncompromising—and it was. The Europeans backed off for the time being. But the BC forest industry was left with a detailed set of laws defining how forestry must be practised. From a silvicultural point of view, the code is comparable to improving the horticultural standards of weekend gardeners through a complex set of rules and regulations about rose gardening, imposed at gunpoint by a corps of "garden police." In seeking to protect our export markets, we ended up with a rigid silvicultural orthodoxy.

In mid-1993 the government passed the Forest Land Reserve Act, to define the forest land base and restrict its use to commercial forestry. A form of land zoning, this legislation was intended to provide industry with an assured land base. Ultimately all government-owned forest land not reserved for parks, wilderness areas or other protected status was to be placed in the reserve. The legislation also included private forest land managed under a state-approved plan in exchange for modest tax breaks. Unmanaged private forest land, i.e. neglected private forests, was not included. Although most managed private land is owned by forest companies, there are several hundred small private forest owners. This category of managed private land is arguably the best managed forest land in the province, having been tended much more carefully than most government-owned land. The owners were outraged to discover that their lands were to be included in the reserve, and they would be forced to manage their forests under the Forest Practices Code. Further, the Forest Land Reserve Act would not permit them to change the status of their land and remove it from the reserve. As one owner observed: "The government has used the Forest Reserve Act to lock us in the room, and will use the Forest Practices Code to beat us to death." They immediately formed an owners' association and protested vigorously. By early 1998 the government had agreed to a more results-oriented version of the Code to apply to these lands.

Meanwhile, other government departments were busy implementing new policies that would also have a significant impact on the forest sector. Foremost among these were developments in Native land claim negotiations. A BC Court of Appeal decision in 1993 determined, in effect, that aboriginal rights must not be unjustifiably infringed upon by activities associated with the harvesting and management of Crown forests. Since the precise nature of those rights has yet to be determined in treaty negotiations between the Canadian and BC governments and tribal councils throughout the province, the Ministry of Forests has entered into a number of interim agreements with several Native bands regarding the use of forests within land claim boundaries. These agreements have also had the effect of reducing allowable annual cut levels in some Timber Supply Areas.

It soon became apparent that the government's new measures might dramatically affect allowable annual cut levels throughout the province. The concerns of forest companies and their workers turned to alarm in early 1993, prompting 15,000 to 20,000 angry forest workers to stage a demonstration against the new policies at the legislative buildings in Victoria. At the same time, the government had to deal with another situation: lumber prices had risen sharply over the past few months; under the terms of the softwood lumber export agreement with the US, BC stumpage charges would have to be raised considerably. The government would find little support among workers for raising stumpage fees at the same time it was reducing AACs, which was causing mill closures and layoffs.

The government relegated these problems to a newly created body, the Forest Sector Strategy Committee, which comprised two groups: a committee of deputy ministers and other senior bureaucrats; and an advisory committee, appointed in April 1993, of fifteen senior industry and union leaders, token Native, environmental and academic members and the University of BC dean of forestry. The primary purpose of the group was to develop an industrial strategy for the province's working forest. Shortly after it was established, a subcommittee with a similar membership profile, the Industrial Structure Working Group, was organized. As its name suggests, it was intended to deal with the structure of the forest sector, including the question of forest tenure.

This approach, with the broad policy mandates assigned to these groups, was a marked departure from previous policy-making practices in BC. The province's four Royal Commissions and one "permanent" commission on forestry were all conventional liberal democratic methods of making policy in a pluralistic society. All interested individuals and groups could present their views in a public forum and policy advice was tendered to the government, which acted in whatever ways it deemed prudent. In the new approach, a decision-making body consisting almost entirely of government, union and corporate members met privately to develop policy, which was then implemented by government. This model of policy-making is generally known as corporatism. It is the way industrial policy is made in Singapore, Korea, Japan and other Asian countries, and in the past it was used in Germany and Italy.

Under the direction of Deputy Premier Doug McArthur, the Forest Sector Strategy Committee worked under intense pressure to hammer out a plan called Forest Renewal BC (FRBC). The government quickly passed legislation establishing a Crown corporation with that name, and assigned it revenues from increased stumpage fees. FRBC's mandate was to employ forest workers whose jobs were eliminated by the government's other forest sector initiatives and put them to work in a range of jobs—environmental restoration, intensive forest management projects, value-added manufacturing and so on.

In 1996, after Mike Harcourt resigned as premier and was replaced by Glen Clark, a further shift in the policy-making process occurred. By this time it was apparent to industry that it was sailing into a sea of red ink. In spite of record high lumber prices and a booming export market, the costs of conforming to the government's Forest Practices Code, and the higher stumpage fees levied to pay for Forest Renewal BC, were being felt. Further, the Timber Supply Review was leading to reductions in the annual timber harvest in some areas, and the Protected Areas Strategy was removing substantial areas from the working forest, reducing cut levels even more. Having cried wolf on numerous occasions in the past, the industry got little response from media, government and the public when it warned of a looming financial crunch.

In fact, Premier Clark's next forest sector initiative was to announce a new "Jobs and Timber Accord." In fact it was not an accord at all, but a decree that the industry was to produce 22,000 new jobs within five years. The government was now acting in classic corporatist fashion, determining policy, which is then implemented through the corporations and unions. Other interest groups—industrial, environmental and so on—which had been brought to the table during the Harcourt years, were relegated to minor roles and the government dispensed with the pluralistic concept of providing a forum for interest groups and attempting to find a solution which meets various, often conflicting needs.

By early 1998 the forest industry was headed into a severe financial downturn, exacerbated by a decline in Asian markets. Mills were closing at an alarming rate and thousands of workers were being laid off, temporarily and permanently. At the same time, environmental organizations, marginalized in the Clark government's decision-making processes, were mounting a new series of campaigns to restrict logging further in various parts of the province.

By fall the government was forced to admit publicly the industry was in trouble. In response it set up two processes to deal with the problem. The first, known as the thirty-day process, was to find some quick fixes to some of the regulatory provisions that were costing the industry money. The second, called the ninety-day process, was intended to deal with larger policy issues such as tenure.

These processes were placed in the hands of a senior NDP-affiliated bureaucrat, who embarked on yet another policy-making procedure. A few senior bureaucrats met privately with a select group of "stakeholders"—individuals, representatives of industrial sectors and interest groups—to discuss various policy options. No public meetings were held. By mid-summer the process had not produced a plan deemed worthy of acceptance by the government. But yet another policy review process was announced, beginning in late 1999.

Meanwhile the government announced a plan to compensate forest companies for land and harvesting rights that were confiscated for new parks. The plan included a proposal to pay compensation in the form of title to Crown land. A limited means of discussing this radical departure from existing tenure policy was established by setting up a series of small public meetings in a few coastal communities, under the direction of a former NDP candidate. By this time it was clear the government did not have a coherent means of developing new forest policy and was flailing about in a futile attempt to cope with the deteriorating condition of the industry.

There was little evidence that the people of BC accepted the policy-making initiatives that evolved under the Clark administration. This was not good for the forest industry: to maintain and improve its health, the industry must have the confidence of the public. Under the Clark government's system of policy making—or lack of a system—that confidence was not possible. BC is a pluralistic society and requires a public forum, such as that offered by a Royal Commission, to facilitate the formation of new forest policy.

Building the Forest Industry 3

The present British Columbia forest industry is the product of an evolutionary process that began almost 150 years ago. The industry has been shaped not so much by sudden, radical changes in policy, as by gradual shifts in emphasis. It has evolved largely in response to the ever-changing opportunities available in foreign markets, not domestic ones, so it is largely a creature of those markets.

The first Europeans to visit BC found a land covered with magnificent and intimidating forests. Those who arrived by sea were astounded to learn that these forests stretched from northern California to the Russian-controlled territories of Alaska, in dense stands that were almost continuous. The forests were comprised of many unrecognizable species.

Indigenous inhabitants did not make extensive use of this forest. The coastal population obtained most of its food from the sea or the foreshore. The most important tree species was western red cedar, which, because it was easily split and worked, was used for almost everything—its bark for clothing, its timber for housing, utensils, boats and ceremonial carvings. Other tree species were seldom used. Natives were highly dependent on the existence of a large stock of high-quality old-growth cedar, but considering the volume of standing timber, their impact on the forest was insignificant.

On a global level, wood was the most important strategic material of that era. The power of a nation was in direct proportion to the number and quality of the sailing ships that flew under its flag. Naval strength depended on access to timber best suited to building ships. In the early part of the colonial period, much of the world's important timber came from Russia, Sweden, Finland, Norway, Latvia, Estonia and Lithuania. Of particular importance were the coniferous trees—pines and firs—that grew there. The major colonial powers of England, France, Spain, Portugal and Holland could sometimes furnish themselves with the deciduous species used in shipbuilding, mainly oak. But they could not supply the tall, straight coniferous species used for masts and spars. It was the constant search for good spars that lured Europeans into the eastern North American forests and first prompted their curiosity about the forests of the Pacific Northwest.

A good spar must be strong, tall and straight, and resistant to rot. The first sea captains to visit the BC coast were quite knowledgeable about these requirements and skilled in selecting trees to meet them. In 1778 Captain James Cook dropped anchor in Nootka Sound, on Vancouver Island's west coast, looking for suitable material to replace his rotting spars. The shores were lined with Douglas fir trees, and a few weeks later Cook left for England, his ship rigged with new fir spars.

Over the next ten years, the occasional explorer and trader followed Cook's example, and a mild interest in this new and unknown timber began to develop. In 1788 another British sea captain, John Meares, brought a crew of Chinese shipwrights to Nootka Sound and put them to work hewing a deckload of Douglas fir spars, which he intended to sell in China. His cargo was lost at sea, but interest remained strong.

By 1847 a sawmill industry had developed in Washington and Oregon, supplying lumber to the booming California market. From these mills the British navy obtained a shipment of Douglas fir, which was taken to its shipyard at Portsmouth for testing. It was found to be superior even to Latvian timber, then considered the world's best for ships' spars.

In 1860 Edward Stamp, another British captain, was able to persuade a group of fellow countrymen to finance construction of a sawmill on Vancouver Island. The Anderson mill, at the head of Alberni Inlet, was BC's first mill built to serve the export market, which at the time was volatile and unpredictable. The Civil War had disrupted export industries in the United States, while demand for timber was growing throughout the Pacific and in Europe. The US imposed a tariff of $1 per thousand board feet on lumber imports, which encouraged Stamp to sell into offshore markets. The colonial authorities, anxious to help him compete with the Americans, provided him with timber for almost nothing. It was not the last time BC's timber pricing policies would be influenced more strongly by outside circumstances than by supply and demand within the province. This fundamental difference between the way timber is priced here and in the US, where most home-grown timber is used domestically, has been a source of conflict and disagreement ever since.

Prior to this, in 1848, the Hudson's Bay Company had built the province's first water-powered sawmill at Victoria to supply local needs, and a year or two later Captain Walter Grant built a mill at Sooke, which in 1860 was rebuilt and equipped with steam power. In 1858 the Land and Fleming mill began cutting lumber at Yale; it sold the wood to the first wave of gold rush miners for $125 a thousand board feet.

In the same year the Anderson mill opened, R.P. Baylor was operating a sawmill at Antler to feed the domestic lumber market created by the Cariboo gold rush. At about the same time another mill was operating on the Wild Horse River, near the present site of Fort Steele, where other gold rush activity was taking place.

The Anderson mill closed after four years, having cut only 35 million board feet of lumber. Because oxen were used to skid logs from the woods to the water, where they were floated to the mill, the mill soon ran out of accessible timber even though the Alberni valley was one of the most timber-rich regions in the entire province. That same year, Stamp opened another large export mill in Burrard Inlet; it became known as the Hastings Sawmill. At approximately the same time, another smaller local mill on the north shore of the inlet at Moodyville was taken over by local business interests and expanded to produce lumber for the export market.

By the time BC joined Confederation in 1871, a pattern had been established in the forest industry, most of whose operations took place along the coast. Numerous small, locally owned mills supplied domestic markets, cutting lumber for the rapidly developing urban centres and the growing farm communities in the Fraser Valley and on Vancouver Island. The larger export mills were developed and operated by foreign investors, primarily British.

The mills obtained their timber through Crown land grants and, after 1865, the first of the temporary tenures giving rights to cut timber only. Contractors such as Jeremiah Rogers, who worked for Stamp at Alberni and then in English Bay, logged for the larger mills. The typical contractor employed a few dozen workers, who used oxen to drag logs along prepared skid roads. This production was supplemented by handloggers, who cut timber along steep shorelines, used their own muscle power to slide the logs into the water, then sold them on an open market to both domestic mills and export customers. Thus, from the outset, both contract and independent logging outfits operated on the coast.

By the late 1880s some new operations were established. The first of what would become a flood of American investors purchased a large tract of Esquimalt & Nanaimo (E&N) Railway land on eastern Vancouver Island and built the Victoria Lumber and Manufacturing Company (VL&M) export mill at Chemainus in 1889. The mill was financed by a syndicate of US lumbermen, including Frederick Weyerhaeuser, under president John Humbird. Most American lumbermen began their careers in the forests of Michigan and Wisconsin, moved west as the timber was cut and ended up in the Pacific Northwest.

Construction of the Canadian Pacific Railway in 1885 launched another phase of the lumber industry. The building of the railway itself—its bridges, snowsheds, roundhouses, stations, warehouses and other facilities—consumed enormous quantities of lumber. Settlement of the prairies created a booming market, supplied by sawmills built in southeastern BC. Most of these mills, which cut their timber in the Interior wet belt forests of the Kootenays, were financed and operated by eastern Canadian capital.

By the turn of the century there were two distinct markets, each served by its own sets of mills. The cargo market operated on the coast, where the large export mills cut lumber and delivered it by ship to customers in the UK, China, Australia and elsewhere. The rail market, located in the Canadian prairies and to a lesser extent in the US, was serviced by many of the Vancouver-area mills as well as by mills in Interior centres such as Fernie, Nelson, Cranbrook, Creston, Revelstoke and Golden.

During the first decade of this century, American timber speculators arrived en masse, followed soon after by US investors who established logging and milling operations. The forests in the eastern US had been largely cut over by this time and the timber speculators had moved west, only to be frustrated when the conservationist-inspired Roosevelt government locked up huge areas to protect future timber supplies. Encouraged by the McBride government's new leases, they swarmed into BC. The industry that developed in their wake was largely American controlled, as was the growing lumber export brokerage business that operated from San Francisco, Portland and Seattle to sell US and BC lumber abroad.

BC's lumber and logging industries flourished during this period, up to the start of World War I. In 1909 the forest sector underwent a further diversification with the opening of two coastal pulp mills, one at Swanson Bay on the central coast and the second at Port Mellon in Howe Sound, which included a paper mill. An earlier paper mill, built in Alberni Inlet in 1894, had used rags—mostly old sails—as a feedstock, and failed after two years. In 1912 two more pulp and paper mills opened, one at Powell River and another on Howe Sound. That same year a small paper mill was built on the Saanich Peninsula to manufacture asphalt roofing materials. As well, the province's first plywood plant opened in 1913 at Fraser Mills, upriver from New Westminster.

Another development reflecting the growing importance of the forest industry was the creation of the provincial Forest Service—first called the Forest Branch—in 1912. The new agency took on the tasks of protecting forests, overseeing and collecting revenues from timber harvesting operations and aiding in the development of the forest industry. At this time, with the Panama Canal about to open and take 13,000 kilometres off the sea voyage to Europe, the BC economy was booming, fuelled primarily by a rapidly expanding forest industry.

Then the roof fell in. First, in 1913, the prairie market withered with a decline in immigration. The following year, when war broke out, the domestic market was thrown into confusion, and the export market collapsed when US brokers stopped loading BC lumber on their ships for fear they would be sunk by German naval vessels.

A year later, with the provincial economy in tatters, the province's new chief forester, H.R. MacMillan, was dispatched on a global tour to sell BC lumber. He visited England, Europe, South Africa, India and Australia, sending home single orders that surpassed the total export volumes of previous years. In addition, he arranged shipping through the Royal Navy. He carefully studied the timber needs of the countries he visited and established a string of contacts among their lumber dealers.

Upon his return to Victoria, MacMillan attempted to persuade the big sawmill owners to form a co-operative marketing organization to sell their lumber abroad, bypassing US brokers. The nature of the lumber export business at that time was such that it was difficult, if not impossible, for a single mill to engage in the export trade on its own. Typically, large overseas dealers placed orders for shipload lots that no single mill could fill in the time available. Consequently, an intermediary was needed to accept orders, purchase lumber from several mills, and assemble it for shipment. This role had previously been performed by the US brokers and, as MacMillan realized, would have to be undertaken locally for BC mills to succeed in the export business. For the duration of the war, the Forest Service performed this function under the direction of W.J. VanDusen, a young forester MacMillan had lured west from Toronto.

After the war, with the established mills still declining to work together, MacMillan, who had left the Forest Service, set up a timber export company, the H.R. MacMillan Export Company, with VanDusen and a prominent British dealer as partners. The company accepted orders from abroad and filled them with lumber purchased from the larger BC mills, arranging for assembly, shipping and insurance. The venture was enormously successful from the outset, partly because MacMillan had promoted Douglas fir in the UK. Until the war, most Douglas fir lumber exported was high-grade, clear wood. To obtain this, the mills had to produce large volumes of lower-grade lumber, which was sold domestically at very low prices. Recognizing the constraints this practice placed on export potential, MacMillan set about persuading his established overseas customers to use this lower-grade fir for other purposes, such as railway ties and construction timbers.

The 1920s was a decade of prolonged growth and expansion, fuelled by strong global and local markets and the increased availability of logging and milling technology designed to process the big timber that grew in the Pacific Northwest.

During the post-World War I period, the industry matured rapidly. Logging operations were mechanized, with steam yarders replacing the remaining oxen and horses. The practice of progressive clearcut logging was born: steam-powered railways were pushed far up the valleys, which were stripped of timber from one end to the other. The concept of forest management, which had emerged around the time of the Fulton Commission and the formation of the Forest Service, was forgotten in the assault on the province's forests.

Various sectors of the industry became clearly defined and organized into industrial groups such as the BC Loggers' Association, the BC Lumber Manufacturers' Association, the Plywood Manufacturers' Association, and the BC division of the Canadian Pulp and Paper Association. The coastal industry was characterized by the development of numerous large firms, which used railway-based logging systems and often operated their own sawmills. At this stage there was not a great deal of integration within firms. The pulp mills were commonly independent operations, without sawmills or other conversion plants. Their timber was obtained from Pulp Leases, which consisted of large areas of low-quality timber and smaller amounts of high-value logs, which could be traded with sawmill companies. A large number of small logging companies sold their timber into the Vancouver log market, where it was bought by the numerous independent mills that did not operate their own logging camps.

In the Interior, forest-based industrial activity was pretty much confined to the railway corridors and some of the large lakes that provided access to rail lines. A few small mills served local markets in areas without rail transportation. The primary species utilized was white pine and, to a lesser extent, Douglas fir. Lodgepole pine, the dominant Interior species, was considered a weed and not used at all.

The Depression slowed the industry, but not dramatically. Provincial log production bottomed out in 1932 at about one-half its 1929 level, and then rose steadily. One of the major consequences of the economic decline was its impact on the organization of the industry. In 1919 the coastal export sawmills had banded together into the Association of Timber Exporting Companies (ASTEXO). The association functioned quietly until 1935, when some of the mills formed their own sales co-operative, Seaboard Lumber Sales. They refused to sell to MacMillan, who had been selling their lumber successfully. A bitter rivalry sprang up between the two groups, with the large coastal mills selling through Seaboard and MacMillan purchasing lumber from smaller mills, as well as entering into the sawmill business himself.

The organized labour component also matured during the 1930s. Woods workers had begun to organize before World War One, after the Industrial Workers of the World—known as the Wobblies—were established in 1905. As much a radical left-wing political movement as a labour union, the Wobblies exerted considerable influence among the unorganized workers in coastal BC logging camps. The first real woods union, the BC Loggers' Union, was formed in 1918. A year later it was folded into the much larger Lumber Workers' Industrial Union, which staged a series of strikes and walkouts through the early 1920s. These actions met with little success and by 1926 the union had virtually collapsed.

Attempts to resuscitate a loggers' union through the late 1920s and early 1930s culminated in a bitter strike in 1933, when the Lumber Workers' Union shut down almost every coastal camp. The LWU went on to organize substantial numbers of coastal millworkers before joining up with unions in Washington and Oregon to form the International Woodworkers of America (IWA) in 1937.

Ten more years and another world war would come and go before the IWA would gain even a modicum of acceptance as a legitimate part of the industry. Until that time it was treated as a pariah, its organizers chased out of camps and mills and its leadership shunned by the captains of industry, who considered it part of a global communist conspiracy—which, to some extent, it was. In the late 1930s, the real power in the industry was lodged in the associations. The BC Loggers' Association ran the woods and, after 1933, led the fight against the union. Two rival associations reflected the intense competition in the lumber sector: the Western Lumber Manufacturers' Association speaking for the mills affiliated with MacMillan, and the BC Lumber Manufacturers' Association representing the Seaboard group. These interest groups were accepted by government and the public as authentic representatives of legitimate interests. Unions had not yet achieved that status, even among workers.

Industry and government had few dealings with each other. Most BC timber was being cut from Crown grant lands or McBride timber licences. The only timber rights being issued by the government were competitive Timber Sales, most of them to small logging firms. Consequently the Forest Service played a minor role in the industry and provincial forest policy was not a significant factor in its operations. In any case, the conservationist vision of forest management was dormant. Forestry meant logging and few people thought about the future. Provincial forest services across the country had degenerated into bureaucratic organizations devoted primarily to their own perpetuation. They collected revenues for government, kept timber theft to a minimum and provided some protection services. Beyond that, they had no important role and little influence on the industry.

During the early 1930s, the BC Forest Service made its first direct venture into forest management with a few small tree planting projects on logged-off land on Vancouver Island. In 1938 an enormous fire near Campbell River burned thousands of acres of the most productive old-growth and new forests in the province. The Forest Service mounted a major effort to reforest this burned-off area, building nurseries and organizing crews of tree planters.

In the pre-war period, industry initiated as many forest management practices as government. Some large Vancouver Island companies that owned land in the E&N belt began to plan and work for subsequent timber crops. Bloedel, Stewart & Welch, for example, replanted logged-off lands. MacMillan, who had acquired his own forest land, limited the size of clearcuts to facilitate natural regeneration.

A series of technical developments that matured during the Depression years introduced a new factor into the logging industry. Trucks, powered by internal combustion engines, were adapted to haul logs and began to appear at various points along the BC coast. Among their virtues were their low cost, relative to a railway system, which created new opportunities for small business firms that could obtain timber through Timber Sales. With trucks, they could afford to log much smaller stands of timber, and they could operate in steeper, rougher areas, where railways could not be built. The use of trucks grew rapidly during the late 1930s with the arrival of the bulldozer, which could build truck roads, as well as skidding logs to roadside. By the end of the decade there were numerous small, independent market loggers, each employing twenty-five to thirty men and working with a few trucks and a couple of Caterpillar tractors to build roads and skid logs. These companies became known as truck loggers.

In the fall of 1939, with the outbreak of war in Europe, events far from BC once again threw the industry into turmoil. This time, however, the consequences were much different than twenty-five years earlier. Lumber marketing was now controlled by people with a direct stake in keeping the industry operating. A few months after the war began, MacMillan was summoned to Ottawa by C.D. Howe, minister of munitions and supply in Mackenzie King's wartime cabinet, and placed in charge of Canadian timber production and distribution. Within a few months the industry was fully mobilized for war. Logging camps were reopened, mills were readied for increased production and export shipments were redirected. The industry's primary tasks were to supply the UK with timber and to provide materials for construction of airports and hangars across Canada where Allied pilots would be trained. As other wartime industries geared up, the forest industry was also called upon to supply lumber and plywood for a huge program to house munitions workers across the country.

Almost overnight, the industry was transformed. Old rivalries such as the MacMillan–Seaboard and IWA–BC Loggers' Association conflicts were set aside. A massive effort to log and mill aircraft spruce was launched under the direction of BC Chief Forester E.C. Manning. When a shipping shortage threatened to halt the transport of lumber to the UK, the railways were persuaded to haul it across Canada to Montreal and Halifax, where it was loaded on ships. All of these changes were wrought in a few frenzied months. The industry would never be the same again.

Even before the war ended, various influences were at work that would transform the BC industry in the post-war period. A new chief forester, C.D. Orchard, was laying the groundwork for a major shift in provincial forest policy. Large pools of capital earned by producing war materials were available for investment in the industry. A shortage of workers, combined with the wartime experience of working side by side with mill owners, gave labour organizations a new, respectable status. There was widespread optimism about the post-war economy, and well before the war was over a timber boom developed, fuelled by American buyers purchasing every piece of forest land on the BC coast they could get their hands on. They were joined by established BC companies and a few new ones.

Typical of the period was the injection of new eastern Canadian capital into the forest sector. In his attempts to secure more forest land, MacMillan had set his sights on the extensive holdings of the Victoria Lumber & Manufacturing Company at Chemainus. John Humbird—company president, grandson of its founder and current president of Seaboard—refused to sell to MacMillan, so MacMillan appealed to a wartime acquaintance, E.P. Taylor, who bought VL&M and turned it over to MacMillan. His interest in the BC forest industry aroused, Taylor looked around for further opportunities and, with MacMillan's assistance, bought several coastal sawmill companies, along with their extensive forest holdings. He restructured them as BC Forest Products (BCFP), the archetypical post-war integrated forest company.

The BC government's establishment of Forest Management Licences in 1947 appealed to companies like BCFP. Backed by large amounts of external capital, the company could commit itself to building expensive new mills in exchange for the allocation of huge areas of Crown forest land. These new licences ushered in the era of multinational forest corporations, which flocked to BC, acquired FMLs and reorganized themselves as fully integrated operations, with area-based tenures, large timber supplies, sawmills, pulp mills and overseas marketing operations.

If the coastal industry in the pre-war period was dominated by the lumber sector, the post-war era belonged to the pulp companies. During the war hundreds of new products had been developed, from fabrics to films to plastics, using cellulose as their feedstock. An explosion in the use of paper further fuelled the demand for pulp. Faced with new competition from outside the province, established companies such as Bloedel's and MacMillan's also diversified, building pulp mills to utilize the waste from their sawmills, adding new mills and plywood plants. Fibre forestry had come to BC.

By the early 1960s, with allocation of FMLs slowed to a near standstill, the major coastal companies began to take control of cutting rights available in the Public Sustained Yield Units. Independent loggers were squeezed out, bought out and, in many cases, converted into contractors for the big companies. As the supply of good sawlogs available on the coastal log market declined, so did the number of independent sawmills that depended on them. By 1970 the demise of the independent coastal logger and mill operator was an accomplished fact.

In the Interior a different kind of development took place. Thousands of small, portable bush mills appeared after the war, cutting timber obtained through Timber Sales. They delivered their rough-cut boards to planer mills set up in the larger centres, which dried, finished and marketed the lumber. Gradually, control of the timber supply evolved to the planer mill companies.

This consolidation escalated rapidly with the establishment of Interior pulp mills beginning about 1960. The pulp mills were able to utilize the residues from sawmills and planer mills if the logs were debarked before milling and if the mills could add chippers to their operations. This incentive led to the consolidation of the lumber milling industry and the growth of an Interior pulp and paper industry, using sawmill waste as its main raw material. Unlike the coast, the Interior industry had little vertical integration: most sawmills remained in independent ownership.

Coincidental with the rise of the pulp sector was a gradual shift in lumber production. Until the end of World War II, sawmills typically used only high-grade logs and cut from them a range of products, from fine finishing woods to rough construction-grade stock. As high-grade timber became more scarce, as coastal mills began utilizing smaller logs, and as the Interior sawmill industry modernized and began using smaller Interior species, mill output changed. It took a couple of decades, but by the 1970s BC was predominantly a producer of commodity lumber. Particularly in the Interior, new sawmills were equipped with the most sophisticated milling equipment in the world, making the province one of the most efficient commodity lumber producers. By 1984 the BC forest industry utilized less labour and added less value per unit of production than most countries in the world.

From the producers' perspective, this was an admirable accomplishment that enabled them to compete handily in world markets. For the provincial economy it was a different matter. If BC's forest products had been consumed within the province, the overall economy might have benefitted by having access to such low-cost materials. But since most production was exported, the potential value of BC timber was being exported as well. Less value was being added to the trees harvested and fewer jobs were provided. Finally, there were limited opportunities for industrial expansion in this approach: as industry made full use of the operable timber supply, there were few ways to increase production. Apart from buying out competitors—a much exercised option—the only choices were to improve utilization of the existing timber supply and/or begin producing higher-value products. By the end of the 1980s, the industry was coming under heavy criticism for the low value added to its production. The NDP government made value-added production and an increase in forest sector jobs one of its priorities, occasionally at the expense of efficiency and competitiveness.

As other business associations proliferated, the small, independent coastal logging firms formed their own organization in 1942. The Truck Loggers' Association (TLA) was set up to represent market loggers who were experiencing difficulties with administration of wartime purchasing regulations. They had observed the success of the BC Loggers' Association in representing larger companies and decided to emulate this success, rather than join the larger group, where their concerns would be overwhelmed by those of the big companies. Fortuitously, the TLA came together just before FMLs were introduced and was able to organize opposition to the new tenure. The TLA was not successful in stopping the FML, but from 1947 on, it provided one of the few coherent voices opposing the industry's domination by integrated multinational corporations. For at least four decades the association would offer a powerful voice of criticism and a forum for all points of view on provincial forest policy.

The diverse character of the Interior industry in the immediate post-war period delayed allocation of FMLs there. With a few exceptions, FMLs awarded by the late 1950s had been on the coast. When the scandal surrounding the bribing of Forest Minister Sommers had subsided and a new minister, Ray Williston, took office, a new form of tenure was used to develop the Interior forest sector. Williston provided Pulpwood Harvesting Agreements (PHA) to new pulp mill operators, guaranteeing them access to trees too small for the existing sawmills if they were unable to obtain sufficient residues from the sawmills. Although it was rarely admitted, this measure also helped prevent pulp mills from being held hostage by sawmills. In the end, very little timber was harvested under the PHAs because the rapid expansion of the sawmill sector and the development of new technology—primarily the chip-and-saw—provided enough residues for the pulp mills. One consequence was that in the Interior the forest industry was dominated by a sawmill sector with more diverse ownership, and pulp and paper played a secondary role.

The difference between these two sectors was reflected in the labour movement. During the late 1940s and early 1950s, the woodworkers' union underwent a great internal struggle, during which its radical leftist faction was purged and the union emerged as a more mainstream labour organization. In 1963 two new unions, the Pulp, Paper and Woodworkers of Canada and the Canadian Paperworkers Union, were formed to represent pulp and paper mill workers. The long-held labour goal of "one union in wood" was defeated, leaving the IWA with the woods and the sawmills to organize. They concentrated on the sawmills, which were easier to organize than logging operations, as they had larger concentrations of workers and more fixed locations. This was particularly true in the Interior, where the consolidation of Interior mills led to large concentrations of millworkers, but as cutting rights were acquired by bigger companies, logging in the Interior was gradually taken over by small contracting firms with fewer workers, and were less likely to be organized by the union.

Executives of some of the big forest companies that dominated the BC forest industry attempted to turn them into broad-based multinational corporations by diversification. Jack Clyne, who became chairman of MacMillan Bloedel in 1959, was one of them.[1] The company made investments in a number of areas, including real estate development, pharmaceuticals and general shipping. The latter ended in a monumental financial disaster, almost dragging the parent company down with it. After a decade or so of such activity most companies, including MacMillan Bloedel, pulled in their horns and divested themselves of their non-forest interests.

Beginning in the 1960s, a massive reforestation effort spawned a new division in the forest sector, a silvicultural industry. Because of BC's tenure policies, a stable, permanent silvicultural capability did not evolve easily. Instead, small contracting companies hired temporary labour to fulfill short-term contracts with licensees and the Forest Service. Through the 1990s, with tree planting performed through much of the year, stable, full-time work became possible for a portion of the silvicultural workforce. Considering the importance of silvicultural capability to the future of the industry, the exploitative, short-sighted treatment of the BC silviculture industry has been one of the more obvious forest policy failures since the 1960s.

Stability in the silvicultural sector is provided mostly by contractors who formed the Western Silvicultural Contractors Association in 1982. Their seasonal employees are almost entirely unorganized and therefore have virtually no voice in the affairs of the industry. When Forest Renewal BC was established in 1993, it took over the financing of a substantial portion of silvicultural work. Because of its mandate to employ displaced loggers and millworkers—almost all of them union members—many silvicultural contractors found their work disrupted or terminated. By 1997 many of them had gone out of business, experienced workers had lost their jobs and the ever-precarious silviculture sector had been further undermined.

Late in 1997, the government established yet another agency to regulate the silviculture sector, New Forest Opportunities (NewFo). It was to function as a union-style hiring hall for FRBC-funded silviculture. All silviculture workers would be hired by NewFo, and would be forced to join the IWA. Contractors would be required to hire through NewFo.

During its first year of operations, the NewFo hiring system was applied to coastal forest enhancement work, with plans to extend it into the Interior and coastal reforestation employment later. Within a year it was apparent, even to the government, that the insertion of

NewFo into the system was drastically increasing silvicultural costs, and the government appeared to be backing off from its plans to expand the agency's sphere of activity.

The 1970s also saw the birth of an environmental movement that often focussed on BC forests. In part, this phenomenon was an outgrowth of the disaffection many young people developed toward mainstream North American society at that time. As well, many more people were engaged in the recreational use of forests. Those who came of age in the 1960s, 1970s and 1980s were less likely to head for the golf course than to don hiking boots and a backpack, launch a canoe, or otherwise get themselves into the back country. They arrived to find an expanding logging industry moving into remote, pristine forests. The conflict was inevitable.

As each area of forest became a centre for confrontation, numerous organizations with various recreational and environmental interests sprang up quickly. Some represented clearly defined forest-user groups, such as wilderness guides and outfitters, hunters or sport fishermen. Small, local forest use conflicts spawned scores of grassroots groups of resident volunteers. Others, such as the Sierra Club, Greenpeace and the Western Canada Wilderness Committee, gained wider support through their skilled fundraising and media campaigns and were broader in their outlook. They attracted a primarily urban support base and very quickly became a powerful force in the forest sector, something the established players in the industry failed to recognize until they began losing their right to log the timber allocated to them by compliant governments. The Tsitika valley, Moresby Island, Meares Island, the Stein valley, Carmanah valley, the Valhallas, Clayoquot Sound, the Kitlope valley and many smaller regions were targetted by environmental groups. To industry the list seemed endless, and it retaliated with its own well-funded, but often misguided, media campaigns.

By 1991 the industry and environmentalists were squared off in a series of running battles over an increasing portion of the forest land base. The newly elected NDP government attempted to resolve the forest-use conflict by creating a transparent land use planning process and establishing scores of new parks and protected areas. Tensions rose unabated, however, climaxing in a year-long protest along a logging road into Clayoquot Sound that attracted thousands of protesters, some from other countries, and precipitated the largest mass arrest of protesters in BC history. In 1994 Premier Mike Harcourt stood up and announced the achievement of peace in the woods—somewhat prematurely, it turned out. With the election of his successor, Glen Clark, environmental organizations resumed their campaigns with the primary objective of eliminating all clearcut logging in the province, as well as ending logging in old-growth forests. By the summer of 1997, the ground had shifted again. The international wing of Greenpeace attempted to blockade logging operations on the mid-coast and, essentially, were run out of the province on a wave of public indignation.

The cozy relationship between government and the forest industry, which had prevailed since the beginning of the century, had changed before the NDP came back into power in 1991. In 1986, when Bill Vander Zalm replaced Bill Bennett as leader of the Social Credit party and premier of BC, the power structure within the forest sector was badly shaken. Vander Zalm won the party leadership by promising to get rid of what he called "the machine," a back-room group clustered around the premier's office, then occupied by Bill Bennett. The party membership was dominated by small business owners from hinterland communities who were fed up with the close alliance that had been forged between the Bennett government and big corporations, particularly those dominating the forest sector. They gave Vander Zalm a resounding victory.

By this time Mike Apsey, the Bennett-era deputy minister of forests, had gone back to the Council of Forest Industries (COFI) as its new president, severing the industry's direct pipeline to power. Next, Vander Zalm infuriated the big forest corporations with his choice of Jack Kempf, a political maverick with a background in the forest industry, as forest minister. Threats to cut off a forest industry-funded election war chest failed to bring Vander Zalm into line. The "machine" was turfed out and some of its former members, including

Bennett's former principal aide, Patrick Kinsella, moved into a public relations firm, which was awarded a $2 million contract by COFI. It launched a major advertising campaign, "Forests Forever," aimed at improving the public image of the major forest companies. This campaign was one of the most disastrous failures in the history of Canadian advertising. It succeeded only in raising a widespread public suspicion that the industry had something to hide.

The gulf between the premier and the major forest companies widened when Vander Zalm and Kempf refused to go along with the COFI hard-line stand against the latest American attempt to impose a countervail tariff on Canadian lumber. Instead, Vander Zalm agreed to raise stumpage rates. Kempf was eventually sacked for allegedly mishandling his expense account, and Vander Zalm was toppled by a party coup, engineered in large part by principals of COFI's public relations firm. One of these, David McPhee, was installed as the new premier's executive assistant. By then the party was a shambles, COFI was confronting a series of problems of its own making and the old alliance of convenience between government and the large forest corporations had come unstuck.

By this time many member companies of COFI were getting nervous, thanks to the disastrous Forests Forever campaign, the mishandling of the tariff fight and various other misadventures sponsored by the organization. In 1989 the council was taken to court by one of its members: John Brink, a Prince George lumber remanufacturer, accused COFI of being part of a fraudulent lumber grading scheme. His charges were upheld in court and he was awarded damages. The fortunes of COFI declined and several companies threatened to cancel their membership. A new organization, the Forest Alliance, was financed by some prominent COFI members to take over the industry's public relations work, and COFI's budget and staffing were reduced drastically.

In 1994, twenty-four years of discreet government funding of COFI ended, after a decision to terminate the arrangement had been made by Vander Zalm—with few people in industry even aware that it had ever existed. This quasi-secret program began in 1970 as the Co-operative Overseas Market Development Program (COMDP), having grown out of discussions between COFI staff and officials in the federal Department of Industry, Trade and Commerce. It was proposed that an industry marketing program be jointly funded by Ottawa, Victoria and the industry. BC's premier and finance minister at the time, W.A.C. Bennett, refused to go along with the scheme, so Forest Minister Williston and his nervous chief forester, John Stokes, agreed to provide funding, almost $500,000 a year, through the back door in the form of a stumpage rebate to participating COFI members. The source of the money was kept secret, listed in the COFI budget as a "market development allowance," and no provincial representative was appointed to the COMDP steering committee.

When the NDP came to power in 1972, much to COFI's surprise, funding was continued and Sandy Peel (who later served as chairman of the Forest Resources Commission) became the province's representative on the steering committee. Although the government funding programs were no longer a secret, few people knew about them. The grants were raised substantially during the recession of the 1980s, and by the time the arrangement was terminated the provincial and federal governments were providing about half of COFI's $14-million annual budget. After COFI underwent a major reorganization in 1993, many of its functions in the Interior were taken over by regional associations, and a new group, the Coast Forest and Lumber Association, was set up to assume many of its responsibilities on the coast.

COFI's internal changes and the shift in its status reflected other adjustments being made in the forest sector. The old arrangements, in which a few large forest corporations ran the forest sector to suit their short-term interests, were finished. And so were many of the other arrangements that supported and reflected that power structure. As a consequence, the divided industry was unable to act decisively in the late 1990s when it was confronted with yet another US tariff action, renewed environmental attacks and the excesses of the Clark government.

One of the major changes in the industry between 1980 and 1995 was in ownership of the few large integrated companies that control

80 to 85 percent of production. In the early 1980s most major companies were owned from outside the province, many of them by US or global forest corporations. Industry insiders predicted that over the following ten to fifteen years many of these companies, having creamed off the easy timber from their licences, would cash in and leave, because BC's tenure arrangements did not encourage them to stick around and manage their licences for future crops. This prediction, known at the time as the exodus theory,[2] turned out to be devastatingly accurate.

By 1995, at least half of the major coastal companies, including BC Forest Products, Crown Zellerbach, Rayonier and Tahsis, had ceased operations in BC. MacMillan Bloedel was a much reduced version of its former self and had shed its eastern industrial shareholders. Within another four years it sold its pulp and paper operations, and the remaining assets of the company were being absorbed by Weyerhaeuser. By 1997, Repap and Avenor were folding up, and Fletcher Challenge had liquidated all but its BC pulp and paper operations, and assigned its timber holdings to a new company, Timberwest. A number of home-grown companies like Slocan, Interfor, Doman, Tolko and Riverside picked over the bones and forged new, large operations. Control of most of the large firms is now mostly in the hands of BC-based firms. They are major forest companies, but most of them would not be the size they are today if their predecessors had not abandoned ship. Local ownership may be desirable in some respects, but this particular turnover also reflects international investors' distaste for the turbulent BC forest industry.

By the late 1980s, the Association of BC Professional Foresters had wandered far from its role as guardian of the forests and had become instead an enforcer of professional orthodoxy. Its code of ethics was rarely, if ever, used to discipline members who were responsible for most of the poor forest practices committed since 1947. The association was, however, diligent in hauling up on the carpet any member with the temerity to criticize a fellow professional, or his employer, for practising bad forestry.

Herb Hammond, an ABCPF member from the Interior with a passion for a more ecologically based form of forestry, became one of these after he wrote a contentious report for the Nisga'a Tribal Council documenting mismanagement of TFL #1 in the Nass valley. Coincidentally, one of the professional foresters closely associated with the management of this TFL was Dave Parker, at that time Vander Zalm's second forest minister. The ABCPF subjected Hammond to a long series of ethics investigations that appeared to be designed more to intimidate him and to deter criticism of the forest company than to uphold professional standards of conduct. The incident precipitated a crisis within the ABCPF.

After Vander Zalm dismissed Parker following his bungled attempt to convert Forest Management Licences into Tree Farm Licences, the ABCPF went through a substantial self-examination. A new generation of younger foresters was elected to head it, and began by conducting a major review of its ethical standards and an overhaul of its disciplinary procedures, including the appointment in 1993 of two non-foresters to the disciplinary committee. Change was clearly in the wind in the ABCPF and elsewhere within the industry.

In 1987, in the wake of the softwood lumber dispute with the US, the IWA severed its relations with the international union and set itself up as an autonomous Canadian labour organization, with the bulk of its membership in BC. Its long-time head, Jack Munro, retired to take over the chairmanship of the Forest Alliance, the pro-industry lobby and publicity group funded largely by the forest companies. One of his closest and most visible associates in this endeavour has been Patrick Moore, one of the original founders of the environmental organization Greenpeace. Ten years previously the idea of these two individuals or, at least, the causes they represented, working in tandem on behalf of the forest industry would have been ludicrous.

Over the previous decade the union had lost close to 20,000 members, mostly because of increased mechanization of mills and woods operations. The IWA's authority in the industry had been undermined, and its pension fund and security of its retired members were compromised. In the 1990s the union began organizing workers outside the forest industry.

The forest unions, and particularly the IWA, recovered some status and authority with the election of the NDP government in 1991. When Forest Renewal BC was established a few years later, the IWA's president-in-waiting, Roger Stanyer, was appointed head of the project, and many other positions in the organization were filled by former IWA staff. By 1997 the agency had in many respects become an adjunct to the union.

By this time the BC forest industry was in bad shape, in spite of some of the best lumber markets in history. Timber harvests in several parts of the province were lower because allowable annual cuts had been reduced. The area of parks and protected areas in BC had been doubled, leading to some further reductions in timber supply. Throughout the industry, adhering to the new Forest Practices Code had forced logging costs up substantially, and stumpage rates had risen dramatically in response to high lumber prices. In its interventionist attempts to resolve some of the real problems of the forest sector, the NDP government had stripped the last bit of fat from industry. The collapse of Asian markets in late 1997 brought the coastal forest industry to its knees, although most of the Interior mills continued to operate, many of them after brief closures. A recovery of sorts occurred in 1999, fuelled by an unprecedented US housing boom. But because the industry had not gone through a basic restructuring, few insiders expected a long-term reversal of fortune.

How the crisis of the late 1990s will be resolved is unclear. Higher lumber and pulp prices will conceal some of the inherent weaknesses of the industrial structure and draw attention away from the need for new policies. Even so, the chickens hatched by decades of dishonest and misguided government actions have come home to roost. No doubt future governments will seek to rectify some of the bureaucratic excesses indulged in by the NDP in its attempt to deal with the consequences of the underhanded shenanigans visited upon the province by the Social Credit governments of the 1970s and 1980s. What is clear is that many more changes in the way the industry operates are necessary before it can once again look ahead to a secure, prosperous future.

The Forest Economy 4

The forest economy of British Columbia is entering its third stage. The first, which began in the 1860s, was a purely exploitative one, concerned only with the jobs, profits and public revenues that could be obtained from the forests in the short term.

The second phase began just after World War II, with the adoption of a regulated harvesting schedule. The objective was to stabilize the industry at the highest sustainable level of activity, or as Chief Justice Gordon Sloan defined it, to provide "a perpetual yield of wood of commercially usable quality from regional areas in yearly or periodic quantities of equal or increasing volume."[1]

The third stage involves cultivating new forests to provide an ongoing industrial timber supply. It parallels a shift from a hunting and gathering economy to an agricultural economy. In some parts of the world, this stage is well advanced. In British Columbia, the transition to an economy of forest farmers has barely begun and its success is by no means assured. It will succeed only if those working in and around the forests undergo fundamental changes in thought and action.

Beyond this lies what could be considered a fourth stage, or perhaps simply a refinement of the third. It entails the management of forests for various values, including timber, and the maintenance of the ecological integrity of the forest. The industry in BC is under growing pressure to operate at this level—known variously as holistic forestry, ecoforestry and sustainable forest management—but the task is difficult because the industry has not yet mastered the third stage.

For more than 130 years the forest industry has dominated economic activity in BC. As late as the 1980s a deputy premier referred to it as a "sunset industry." That official has, thankfully, long since returned to the east from whence he came, and the forest industry continued to thrive. Recently it experienced some of its most active and profitable years ever, and the potential remains strong. However, the economic decline that began in 1997 suggests there are some serious structural problems in the industry, and the future of forestry in BC is still cloudy.

The BC "forest industry" can more accurately be called the "forest products industry," or the "timber industry." It encompasses all endeavours related to harvesting and processing trees into saleable products and commodities. Increasingly, it also includes management or cultivation of forests.

Traditionally the forest industry has been separated into three sectors: solid-wood industries, paper and allied industries, and logging and other woods-based activities. In 1998, the logging sector harvested 68.6 million cubic metres of timber. This volume was down from the record 1987–88 harvest of 89.1 million cubic metres. Still, in 1998, the industry processed those logs and sold the products made from them for $15 billion, down from the record $17.7 billion in 1995. About 80 percent of sales were to customers outside Canada. This undertaking directly employed 99,400 workers who were paid $5.7 billion in wages and benefits (see Table 4A).

Like any statistics or speculations on the forest industry, these figures and their implications are the subject of often vigorous debate. Supporters of industry argue that 50 cents of every dollar earned in BC comes from the forests. This is true in that, since the late 1980s or so, more than half of all goods shipped from the province were forest products, and in the late 1990s they accounted for more than 60 percent of exports. Critics of the industry point out that the forest industry provides only about 6 percent of the jobs in BC. They too are correct, although they seldom mention that in 1998 the average wage in the forest industry was $46,030—considerably more than the average BC wage of $32,200.

The forest industry also generates indirect employment in the transportation, service and support sectors, and through the spending of forest industry wages. Depending on whose cause they are supporting, economists calculate that the industry generates from 1 to 4 indirect jobs for every person working directly in it. If we accept the figure of 2 indirect jobs for every direct job, 274,200 people, or about 15 percent of the BC workforce, were employed by the forest industry in 1998.

Another calculation that puts the BC industry in a larger context is the number of jobs generated relative to the volume of wood harvested. From a simple bean-counter's perspective it is desirable to employ as few persons as possible in processing logs. This is called efficiency of production. From another point of view, it is better to generate as much economic activity as possible out of every cubic metre of available timber; that is, to employ as many workers as possible per volume of wood, as long as they add more value to the products than it costs to employ them.

This is not as simple a task as some people seem to think, particularly forest industry critics. While it is relatively straightforward to make more valuable products than, say, commodity two-by-fours, finding secure markets for these products in the highly competitive global marketplace is not straightforward. In theory we could convert our entire timber supply into violins, employing hundreds of thousands of people doing so. But this does not mean the rest of the world would want to buy the millions of violins we could churn out every year. And even if there were a huge, new market for violins, BC's highly paid workers would have a hard time competing against workers from developing countries in such a labour-intensive business. On the other hand, we can probably create more value from our wood supply than we have in the past.

In 1998, for every 1,000 cubic metres of timber logged, 1.4 direct jobs were generated. A decade earlier, only 1 job was created for every 1,000 cubic metres. The increase was dramatic, but not entirely because the industry was adding value to wood. There was some growth in the value-added wood industry and in the silvicultural workforce, but some of the new jobs were created in the Ministry of Forests to enforce the Forest Practices Code. Industry's administrative staff expanded, and more workers were needed in the woods to implement strict new logging regulations. Compared to other jurisdictions, however, BC ranks low in the employment benefits derived from its timber harvest.

VALUE AND JOBS: A 1984 COMPARISON

Country	Volume logged (million m^3)	Value added/m^3	Jobs/ 1,000 m^3
BC	74.6	$56.21	1.05
Other Canada	86.3	$110.57	2.20
United States	410.0	$173.81	3.55
New Zealand	5.3	$170.88	5.00
Sweden	56.0	$79.49	2.52
Switzerland	7.0	N/A	11.41

Source: "Forest Policy: Rhetoric & Reality," by Ray Travers, in *Touch Wood: BC Forests at the Crossroads*, Ken Drushka et al (Madeira Park BC: Harbour Publishing, 1993), p. 196.

The 1984 figure of one job in BC per 1,000 cubic metres cut had declined substantially from two decades earlier. In 1961, 2 persons were directly employed harvesting and processing the same volume of timber. Because of enormous capital investment in new logging and milling technology, worker productivity had been almost doubled. A trend during this period was to produce more standard commodities and reduce diversity of manufactured products. Mills that had once made the most of each log that came up the jack ladder, cutting it into a wide range of products from mouldings to studs, were retooled to turn out at high speed, and with fewer workers, a narrow range of construction-grade lumber and a large volume of wood chips destined for the pulp mills. The new sawmills, most of them in the Interior, extracted less value from each cubic metre of timber, but did it much more efficiently, at a greater profit and at higher wages for workers. Under this industrial strategy, growth and expansion were achieved by acquiring more timber rights, cutting more wood and employing fewer, more productive workers. The same principles were applied in other sectors. Logging became more mechanized and pulp and paper mills more automated, turning the BC forest industry into an efficient producer of low-value commodities.

TIMBER HARVESTS AND EMPLOYMENT

Year	Harvest (1,000 m^3)	Employees	Logging	Wood	Pulp & Paper	Total
			(employees/1,000m^3)			
1990	78,316	76,845	.23	.51	.24	.98
1989	87,414	81,375	.23	.49	.21	.93
1988	86,807	80,361	.23	.49	.21	.93
1987	90,591	80,536	.23	.47	.19	.89
1986	77,503	74,306	.26	.48	.22	.96
1985	76,868	75,921	.25	.52	.22	.99
1984	74,556	76,920	.28	.52	.23	1.03
1983	71,369	77,688	.28	.57	.24	1.09
1982	56,231	75,138	.29	.72	.33	1.34
1981	61,818	86,848	.32	.75	.33	1.40
1980	74,654	95,518	.33	.67	.29	1.28
1975	50,077	76,380	.35	.77	.40	1.53
1970	54,726	73,999	.34	.70	.31	1.35
1965	43,413	73,421	.43	.93	.33	1.69

Source: Burson-Marsteller, 1993.

For the past few years the pendulum has swung in the other direction. With a restricted timber supply for the foreseeable future, the industry is experiencing a renewed interest in adding value to BC timber.

In 1996, having apparently accepted "value-added" as an economic panacea, the BC government announced it expected the forest industry to create an additional 22,000 jobs on a reduced timber harvest within five years, and said it would reallocate timber supplies to achieve this objective. But the economic decline that hit the industry the following year undercut the plan.

Each of the three forest industry sectors has a distinct character that influences the economic opportunities available to the industry as a whole.

Lumber

In BC the sawmill is the primary production facility for raw logs. About 80 percent of all logs are first processed in a sawmill and about 40 percent of these logs are recovered as lumber. The remainder is used in pulp mills in the form of chips, sawdust and shavings, providing about 80 percent of the raw materials used by these plants.

Sawmills on the coast and in the Interior are markedly different, as their forests are different. Coastal mills were built to handle a wide range of log sizes. As the supply of large-diameter old-growth timber disappears, most of these mills have been phased out or converted to handle smaller logs at higher speeds. But even today, coastal mills cut a wider range of log species and sizes into a more diverse array of higher-priced specialty products for shipment abroad—mostly to Japan and Europe—than do most mills in the Interior.

The Interior sawmilling industry underwent a profound transformation during the 1960s and 1970s. Hundreds of small operations were replaced by a relatively few large, highly sophisticated lumber mills—among the most efficient in the world. These mills were built specifically to cut small, uniform Interior logs into two-by-fours, two-by-sixes and other construction-grade lumber. They use chippers to cut away the wood not wanted for lumber, then send it to pulp mills. The saws operate at very high speeds and spit out lumber at phenomenal rates, with great accuracy. Obtaining the highest grades of lumber from each log is not the highest priority, hence the appellation "spaghetti mills." A remanufacturing industry has developed to recut the lower-grade lumber into higher-value products, but often has difficulty obtaining steady supplies of low-grade lumber from the large mills.

In 1998, the province's sawmills cut 12.9 billion board feet of lumber, more than two-thirds of it at Interior mills. This was down from the sector's historic record cut of 15.9 billion board feet in 1987, a feat unlikely to be surpassed for some time. The log harvest that year was well above what is sustainable under existing management conditions, and since then the allowable cut has been reduced.[2]

The 1998 lumber output was sold for $6.4 billion, down $1.5 billion from the previous year. The difference can be found in two fundamental factors affecting the health of the lumber sector: the US housing market and the strength of the Canadian dollar. The US market currently absorbs about 60 percent of the lumber cut in BC, most of it from Interior mills. Another 20 percent is sold in Canada and the remainder goes to other countries, particularly Japan. When Americans are buying and building houses, the Interior lumber sector thrives. When it slows, even if activity is sustained in other parts of the world, the region suffers. Similarly, when the Asian market falls off, as it did in 1997, the coastal industry suffers. The weaker the Canadian dollar against currencies of the US or other countries buying our forest products, the less it costs them to buy from us, and the more they buy. A drop of 1 cent in the Canadian dollar against the US dollar has been estimated to raise BC forest product revenues by $180 million. For example, lumber revenues in 1994 were $1.2 billion more than the previous year, even though 200 million board feet less was sold. That year the Canadian dollar declined by 5.6 percent, and US housing construction increased by 13 percent. The improvement had little or nothing to do with any performance improvement in BC.

The Achilles heel of the BC lumber industry is its dependence on foreign markets. In fact, BC is the world's largest exporter of softwood lumber: about 9 percent of the softwood lumber cut in the world is milled in BC, and 35 percent of the world's softwood exports come from BC. But the factors influencing the lumber industry most heavily—construction activity in other countries and the strength of the Canadian dollar—are beyond the control of the province. In 1995, when housing starts in the US faltered, demand for lumber from the BC Interior fell dramatically. Prices dropped from just below $500 a thousand board feet to less than $250 in six months.

This situation activated another hazard for BC lumber producers. During the same period that demand for lumber declined, demand for pulp rose dramatically and prices doubled in less than a year. This produced an enormous demand for chips, sawdust and shavings that make up 80 percent of pulp mills' raw material needs. In order to obtain these by-products, to which 60 percent of all logs entering a sawmill are converted, the mills had to continue cutting lumber and feeding it into the softening US market, which drove prices down further. US lumber producers went to their government and requested a countervailing tariff on Canadian lumber imported into the US, and an agreement was reached to place a quota on duty-free lumber entering the US from Canada. The complexities of this issue go far beyond the dynamics of the lumber sector, and will be explored later. But these dynamics are one of the difficulties of operating an export-based commodity lumber industry that is largely dependent on a single market, particularly when that market happens to be the protectionist US.

Employment in the lumber sector during 1998 was 2,000 fewer workers than the previous year, at 21,000. During the year, five large sawmills closed. Lumber production costs in 1998, at about $500 a thousand board feet, were down from the previous two years but substantially higher than the 1992 level of $375, mostly because of higher log costs due to implementation of the Forest Practices Code and higher stumpage costs.

The future of the BC lumber industry is uncertain. Escalating production costs are making it less competitive in its major export markets, the US and Japan. The Japanese market has favoured lumber cut from old-growth coastal timber, which is gradually disappearing, and our second-growth forests will not be able to satisfy that market if we do not take care of them. Japan has a large natural forest of its own. It also has an extensive plantation forest that was established after World War II. Harvesting will begin in this forest by about 2010, and it grows about 76 million cubic metres of timber each year—equivalent to BC's annual harvest. The Japanese government recently announced its goal of increasing the domestic lumber supply from 30 to 46 percent of consumption by 2015 and has committed $1.95 billion (Cdn) annually to achieve this objective. It is uncertain when Japan will recover from the 1997 crash, and whether that recovery will restore BC lumber markets when it does occur.

In the US, an enormous southern pine forest stretching from Houston, Texas, east and north to Washington, DC, is beginning to reach harvestable size. Originally conceived as a pulpwood forest, it is now being eyed by its owners—forest companies and small private landowners—as a sawlog forest. They are considering abandoning planned thirty- to forty-year pulpwood rotations followed by clearcutting and planting, in favour of a series of commercial thinnings over eighty- to one hundred-year rotations. This shift in thinking is reinforced by a growing conviction among environmentalists and people working in the forest industry that the US should become much more self-sufficient in forest products. They believe the US should not supply itself with lumber at the environmental expense of other countries, and they also wish to reduce trade deficits. If their ideas are put into practice, the impact on the BC lumber business will be dramatic. As it is, in the period 1996 to 1998, when BC lumber production fell by 1.5 billion board feet, US production increased by 6 billion feet. During this same period, eastern Canadian production increased by 1.1 billion feet, with most of it sold in the US.

Competition in the US market appeared from a new quarter in 1997. Late that year, Swedish mills began selling high-grade pine lumber in the American midwest, in direct competition with BC producers who had been selling into that market for many years. At about the same time, Swedish, Finnish and Austrian lumber producers obtained the right to use US grading standards for dimensional lumber, enabling them to sell construction-grade lumber in the US for the first time. This is BC's primary market; the prospect of being undersold in it by distant European producers is one more sign of the unstable condition of the BC industry.

Plywood and Veneer

BC's plywood industry was founded on the availability of high-quality Douglas fir, mostly during and after World War II. In recent years production has fallen off drastically and several coastal plywood plants have closed permanently. Production peaked in 1987, when 2.5 billion square feet of three-eighths-inch plywood was turned out by 5,100 workers. In 1998 about 2.2 billion square feet was produced by 2,900 employees. As with lumber, log costs for plywood and veneer mills almost doubled between 1992 and 1998.

The decline in the BC plywood business is usually explained by the difficulty of competing with panel boards produced by gluing flakes or strands of low-quality wood into sheets known generically as oriented strand board (OSB). But there are other factors in the equation. Plywood manufacturing is more labour intensive than lumber production, and plywood is made from logs that can also be converted into high-grade lumber products. It is easy to see why the large forest companies have moved away from plywood production.

Ironically, one of the other drawbacks to producing plywood—the lack of an export market—has recently changed. Many countries that buy other forest products from BC have erected tariffs and prohibitive product standards against imported plywood. Canada does the same thing, for the same reason—protection of the domestic industry. But under the Free Trade Agreement, the 20 percent US tariff on Canadian plywood was eliminated at the beginning of 1998, opening the large US market to Canadian producers—and vice versa.

Plywood is a good example of a value-added product. It sells for several times the price of an equivalent volume of lumber produced from the same logs. In many applications it can compete well with the newly engineered panel boards. Its production generates more employment than conversion of the same logs into lumber, as well as additional revenues. But under the conventional decision-making processes employed in much of the forest industry, it is a loser.

Pulp

Of the 7 million tonnes of wood pulp produced in BC's twenty-four pulp mills in 1998, 4 million tonnes was exported. That year, 3 million tonnes of paper and paper board were produced in the province's twelve mills. Of all the pulp-producing regions in the world, BC mills utilize the smallest proportion of pulp for production of paper.

The reasons have to with the wide variety of grades and types of paper demanded in major markets, and BC's distance from those markets. Apart from newsprint, most papers are made by blending different types of pulp. The pulps produced in BC, while of good quality, are blended with other pulps or non-wood materials to produce the great variety of papers used throughout the world today. It is much less costly to ship pulp from BC to paper mills near those markets than to import the additional materials, make the paper here and ship it around the world.

Until recently, the newsprint business operated on a different basis. Newsprint has traditionally been made entirely from wood pulp and was therefore made here as easily as elsewhere. Most BC newsprint is sold in the US, particularly California, with Canadian and Japanese markets taking most of the remainder. However, a growing demand in some parts of the world for newsprint made with recycled newspapers threatens BC producers, because there are not enough old newspapers in the province to provide an acceptable level of recycled material in BC newsprint. BC pulp production costs, particularly on the coast, are already higher than in any other country in the world, so if recycled newspapers have to be imported, the added costs may have serious consequences for the future of the sector.

DELIVERED COSTS OF MARKET PULP, 1998
(Canadian $/Tonne)

BC Coast	$750
BC Interior	$639
Sweden	$529
Finland	$562
Eastern Canada	$625

Source: Price Waterhouse, *The Forest Industry in British Columbia*, 1998.

One of the realities of the pulp and paper business is an enormous cyclical fluctuation in prices. At the end of 1993, for instance, market pulp prices stood at $390 US per tonne. During the following year, they rose to $700 a tonne, and by mid-1995 had reached a record $1,000 a tonne. Then the market collapsed, and by the end of April 1996, prices were down to $490 a tonne, after which they began to rise again, settling at $500 by the end of 1998.

The largest market for BC pulp in 1995 was western Europe, which used 36 percent of the province's production, while Japan took 21 percent and the US 19 percent. In 1996, however, European purchases fell by 15 percent because of a contraction of the fine paper market, while sales to non-Japanese Asian markets increased by an equivalent volume. Sales of market pulp in 1998, down from the previous year, were $2.3 billion, giving the sector a net loss of $373 million for the year. The last year the pulp sector showed a profit was 1995.

Newsprint sales were up slightly in 1998, at 1.6 million tonnes, and a small $46 million profit was earned. The pulp mills employed 6,900 people in 1998, and another 3,100 worked in newsprint mills—1,200 fewer than two years earlier.

A major challenge for pulp producers over the past decade has been to comply with laws requiring them to eliminate specified pollutants from mill effluents. Almost all of BC's pulp mills had been built to employ bleaching processes which produced various toxic waste substances, so between 1988 and 1993 BC's mills spent $1.5 billion modifying their operations to eliminate the pollutants. Since then they have spent another $.5 billion, and they will need to spend an additional $1 billion to eliminate all organochlorides by 2002. Considering the cost, it is worth questioning whether this goal is a realistic priority. The money could probably be more effectively spent on other measures, but the industry has little choice in the matter. Plants must either do the work or close down due to a combination of government regulations and consumer demand for pollution-free paper products.

The influence of the pulp sector on the province's forests has at times been considered pernicious. H.R. MacMillan once described the impact of pulp and paper people as "like a blight on the prairie wheat . . . They do not understand the forest and they will waste and abuse the very foundation of the company . . . They are instinctive monopolists and are not content to rely upon sawmill chips for their furnish. They want first class, whole logs to pour through chippers at high speed. For that they will misuse the existing forest and cut the new forest before it is ready for a proper harvest."[3]

The pulp industry does not need high-grade logs. It requires large volumes of fibre from a few species of softwood trees—spruce is preferred—that can most easily be converted into pulp. Ideally, from the pulp producers' point of view, forests supplying their needs would be monocultural plantations of fast-growing hybrid species, harvested every few years and quickly replaced by a new crop. Plantation forestry, or an extreme version of industrial forestry, is a natural for the pulp sector, although this development is more likely to occur outside BC, where climate and terrain are more favourable.

In BC most of the first pulp mills were built as part of integrated forest companies, with waste products from solid-wood mills as their primary or only supply of material. Most coastal pulp mills were built with whole-log chipping machines to make use of the large volumes of timber found in old-growth coastal rain forests not suitable for sawmills or plywood plants. In the Interior, although some pulp mills were built with this chipping capability, they almost always operated on the residues of the large Interior sawmilling sector. Thus the forest industry that took shape in the 1950s and '60s set out to make the best use of whatever wood was available from the existing forests, and to replace those forests with new ones that would provide a wide range of high-quality timber. The pulp monster, in this scenario, was to be kept safely locked up in the basement, living off the leavings of the solid-wood mills.

The emphasis has shifted during the past thirty to forty years, and the creature from the basement is often found sitting at the kitchen table, demanding a larger share of logs. When lumber prices are down and pulp prices are up, the whole-log chippers run steadily. The pulp and paper sector has gained much more influence within the forest products industry. As MacMillan predicted, control of the companies moved out of the hands of forest-oriented people, like himself, and into the hands of people who were enamoured of the enormous profits that can be made in the pulp business. Bigger and better whole-log chipping facilities were installed in the coastal pulp mills and, in the early 1990s, more whole-log chippers began to appear at Interior pulp mills. Internal competition for a shrinking log supply began to heat up.

The consequences for BC forests are uncertain at this point. Elsewhere—until recently, for instance, in eastern Canada—domination of the forests by pulp producers has often proved detrimental to the forests as well as the people who work in them. On the other hand, the reforestation of the southeastern US is in large part due to the presence of the pulp and paper industry. The trend in a pulp-oriented forest economy is usually away from diversified forests with high timber values, which support a more labour-intensive industry paying good wages, and toward low-rotation timber cropping for a capital intensive industry. However, this trend may be ending, at least in some areas. In the southern US, pulpwood growers are beginning to manage their forests on longer rotations to produce both pulpwood and sawlogs. In part this is a response to environmentalism, but growers also expect better financial returns as rising lumber prices push up the value of sawlogs. Others predict that pulp production will shift out of northern softwood forests into southern plantations, where delivered log costs will probably be a fraction of the cost of feedstock at a BC pulp mill.

Logging

BC's logging sector has been under siege for almost two decades. The assault on the industry was begun by a few environmental groups, then spread to the general public, which applied pressure on politicians and governments. Wounded by new rules and regulations, hamstrung in some areas by reductions in allowable cut and festooned with a bewildering array of new guidelines and codes of conduct, the logging sector, while often attempting to come to terms with its past careless practices, struggles on.

In 1998 the province's 28,200 loggers delivered to the mills 64.8 million cubic metres of logs, a dramatic decline in volume from the 1988 cut of 88 million cubic metres. Less than a third of this timber was cut on the coast and the remainder in the Interior.

Employment levels in the logging sector fell steadily from the early 1970s until the late 1990s, primarily because of increased mechanization. At that point they began to rise again, as companies attempted to comply with the Forest Practices Code, which specifies in excruciating detail how logging will be done. These regulations, combined with other factors such as higher stumpage rates, raised the cost of logging by about 75 percent between 1992 and 1996, according to a 1997 study prepared for the Ministry of Forests.[4] It cost, on average, $37.17 a cubic metre more to log in 1996 than it had four years previously. The cost of hauling logs from the stump to the truck increased by $6.27 a metre, and building roads to the new standards cost another $4.84 a metre. New and more detailed applications for permits, higher road-building standards and smaller cutblocks requiring more roads, planning and administration relative to volumes logged, were among the reasons for higher costs. During 1998, logging costs declined slightly, due mostly to stumpage reductions.

Total logging costs in BC rose by almost $3 billion a year between 1992 and 1996. Higher stumpage and royalty payments accounted for about 45 percent of the increase. Another 33 percent was attributable to costs of complying with the Forest Practices Code. Other new government regulations were responsible for another 18 percent. Only 4 percent was caused by increased costs under the forest companies' control. The 1998 log harvest resulted in stumpage and royalty payments to the province of $1.4 billion, down slightly from the previous few years, but more than double the 1992 amount. In 1998 the industry and its employees paid various levels of government $3.7 billion, even though forest companies lost more than $1 billion that year (see Table 4B).

Logging is carried out quite differently on the coast than in the Interior, because the character of the forests is different. The gentler terrain in the Interior allows more extensive use of mechanized falling equipment, whereas on the coast most falling is still done by hand with power saws. Rubber-tired skidders are widely used to move logs to roadside in the Interior, while cable systems and, increasingly, aerial systems are used most often to skid logs on coastal sites. In the Interior, hauling logs long distances is costly because they are carried from landing to mill by truck. On the coast, once logs are delivered to the salt water, they can easily and inexpensively be towed to any mill on or near the water. For these reasons, logging costs before stumpage in 1998 were $48 a cubic metre in the Interior, down slightly from the previous year but about 50 percent higher than in 1992.

Coastal loggers, because they often work in isolated locations and require large, expensive machinery, are usually organized into larger production units than those in the Interior. The major companies undertake a higher proportion of their own logging operations on the coast. Coastal contracting companies tend to be larger than their counterparts in the Interior, and undertake several or all phases of the logging cycle, from stump to dump. Until recently, the Interior contracting community was composed of a large number of one- or two-person companies, each owning a single piece of equipment and having a contract for a single phase of the operation. But for the past few years the trend has been toward more stump-to-dump contractors in both regions. In 1998 BC contractors logged about 84 percent of the timber harvested, the rest being cut by company-run crews (see Table 4C). In both regions there are only a handful of independent loggers, cutting timber they have purchased and selling it on the open market.

The conversion of BC loggers from independent, entrepreneurial operators who cut timber they obtained from the government in open competition with others, into little more than sharecroppers working for the few large corporations that now possess the bulk of the provincial timber supply, has been a long, painful process. At the end of World War II there were several hundred independent loggers on the coast, selling logs to scores of equally independent mills. In the Interior, as many as 2,000 small mills logged and cut their own timber into boards, which they then sold to larger, centralized planer mills. Forest policies adopted in the late 1940s had the effect of concentrating control over cutting rights on Crown timber in an ever-decreasing number of large companies, many of them integrated, particularly on the coast.

Provisions in some of the licences, such as Tree Farm Licences, required that a certain portion of logging be done by contractors, which is what many of the independent loggers then became. These contractor clauses were not well enforced until a decade ago, when provisions were added to strengthen the contractors' position. Since that time, their lot has improved somewhat and the proportion of logging they do has increased as well.

The tension between BC's small logging companies and the big corporations for which most of the small companies work, typifies a fundamental political tension in the province. The major conflict in BC is not between left and right, between supporters of a state-controlled economy and supporters of the private sector. The basic and sometimes most divisive struggle is between small, independent and indigenous business firms and the large corporations—some multinational, many owned and operated in distant urban centres—that have gained control of the province's resources. When times are good and log prices are high, the conflict subsides. But when times are tough and the financial screws are tightened, the two groups engage in shouting matches, cancelled contracts, arbitration procedures, lawsuits, treks to Victoria to invoke government aid, and contributions to political parties, all of which promise both sides to support their cause.

BC loggers, like their counterparts in Washington and Oregon, have borne the brunt of the public conflict over environmental issues. Although there have been a few protests at pulp mills, focussed on effluent discharges, millworkers rarely experience direct confrontations with protesters. Through the 1980s North American media tended to portray loggers as an uneducated, ecologically insensitive underclass. Perhaps this was because journalists tend to be members of a technocratic urban elite with no cultural connection to the rural people who still maintain the foundations of the economy. Few urban journalists understand rural culture or forest ecology, a state of mind shared by many who organize and lead environmental organizations. The result has been a social denigration of loggers.

The conflict the logging industry found itself in was, in no small way, exacerbated by loggers ignorant of forest ecology and stubbornly insistent on continuing to log in the same destructive manner as they did a decade or two earlier. In other instances they allowed themselves to be used as propaganda tools by those forest companies concerned with maximizing short-term profits before abandoning the woods altogether. Most loggers knew there were serious problems with the way their business was conducted, but in the polarized climate that developed, they found it a risky business to express their opinions publicly. The industry is still burdened with the consequences of this conflict.

Forestry

The forestry sector has two major components, the technocrats and the silvicultural workers.

A large body of technocrats is employed by the Ministry of Forests and the forest companies to administer the forests. Most of them are professional foresters or forest technicians. The Ministry of Forests, with a staff of 4,400 in 1996—down slightly from a year earlier but much higher than in 1991—is charged under the Ministry of Forests Act with encouraging forest productivity; managing, protecting and conserving forests with particular regard to the economic and social benefits they confer; planning forest use with other government agencies; supporting the timber processing industry; and looking out for the Crown's financial interest in the forests. The MoF performs these functions, for the most part, by enacting and enforcing a vast and growing body of laws, rules, regulations, codes and guidelines.

From the time it was established in 1912 until 1979, the MoF operated under a less comprehensive mandate, with its main tasks being administration of timber harvesting, forest road construction, reforestation and fire protection. Essentially MoF employees were policemen of the woods: they were to ensure forests were not being mistreated or, at least, to ensure they were being used only in ways authorized by law. In recent years, the MoF has been assigned several new functions, including a much greater responsibility for managing forests, integrating timber objectives with those of other forest users, bringing about sustainable forest development, and many other vaguely defined tasks.

After its formation in 1912, the Forest Service was built around a highly decentralized ranger system. Under this system the province was divided into a few forest districts, each of which was divided into several Ranger Districts. The forest rangers who were in charge of each of these districts, usually with the help of an assistant or two, were the ultimate authority in interpreting and administering the Forest Act in their area. They were responsible and accountable, through ranger supervisors and district foresters, to Victoria. There were clear, simple lines of authority and administration, and the system was considered effective and generally satisfactory. Loggers with questions got answers right away—the buck stopped at the ranger's desk. There was also little movement between districts and rangers became important members of the communities where they lived and worked.

This began to shift under Chief Forester C.D. Orchard. Well known for his autocratic manner, Orchard established a rigid hierarchy in the Forest Service, reserving all important decisions to himself. He appeared to be in full command of ministers in the Coalition government and, with the advent of Social Credit, tried the same approach, even appointing his own successor when he retired in the mid-1950s. Forest Minister Ray Williston nullified that appointment and restored the chief forester's position to one of equal stature with other senior members of the Forest Service.

The ranger system ran into difficulties in the late 1960s when many new functions and responsibilities were thrust upon the Forest Service, particularly to do with forest management. A centralizing tendency emerged in response, and it was in conflict with the established ranger system. Peter Pearse noted these contradictions in his Royal Commission report of 1975 and recommended an independent review of the Forest Service to "facilitate on-site field responsibility and authority."

Instead, as part of the sweeping legislative changes made in 1978, a Ministry of Forests Act was passed, redefining the responsibilities and mandate of the Forest Service, and opening the way for a major reorganization. The ranger system was abolished outright, the Ranger Districts eliminated, and the province rejigged into six forest regions, which were further divided into forty-six districts. These districts, located in regional centres, were relatively far removed from the field and placed under the direction of a district manager, who sat on top of a hierarchy of functionaries responsible for various activities of the Forest Service, including timber harvesting, silviculture, protection and recreation. Generally, rangers did not become district managers; that position now went to more academically endowed members of the bureaucracy.

Putting the new system into effect was another matter. Many staff were required to populate the new organization, and a vast network of ranger stations throughout the province had to be disbanded so that staff could be relocated in large, new quarters in bigger population centres. Costs soared, and within a few years the government realized it was creating a monster, an assessment shared by most people in industry who had to deal with the new bureaucracy. A freeze was put on hiring; then, as part of the so-called "restraint program" of 1983, hundreds of Forest Service personnel were laid off. As well as undermining morale, this decision destroyed whatever capability the new centralized organization had to function effectively.

At about the same time, public indignation over the industry's harvesting operations reached a boiling point, and people began organizing themselves into protest groups that descended upon the Forest Service, demanding information, action and the right to be consulted on resource planning. Forced by new responsibilities outlined in the Ministry of Forests Act to take into account non-timber forest values, the Forest Service attempted to respond to these new demands. Already in a demoralized, confused state, it stumbled badly and never did recover. Consequently it was unable to fulfill its new role adequately and it contributed enormously to the intense resource-use conflicts that erupted during the 1980s. When the NDP government of Mike Harcourt came along, it quickly recognized the Forest Service's inability to mediate the conflict and began setting up other processes outside the Forest Service, such as the Commission on Resources and the Environment (CORE) and the Protected Areas Strategy (PAS), to deal with the situation. Then, with the establishment of Forest Renewal BC, the Forest Service was relieved of many of its silvicultural responsibilities as well.

Still demoralized, still unable to manage all the responsibilities loaded upon it, the BC Forest Service carries on as a bureaucratic burden on the forest industry. With the adoption of the Forest Practices Code, the Forest Service assumed responsibility for its monitoring and enforcement, hiring between 400 and 500 new, inexperienced staff. The results have been appalling. Implementing the code was to have cost a predicted $5 a cubic metre; under the Forest Service's administration it cost more than $12 a cubic metre. Lacking the resources and field-based organization needed to be effective, the Forest Service has, in some respects, become an albatross around the neck of the provincial forest economy. It is full of good, dedicated people—many of whom, unable to do their jobs, eventually leave in frustration. One of the most compelling arguments for a major restructuring of the forest sector is to redefine the Forest Service's role so it is an efficient, productive player in the forest economy.

As a result of 1948 changes to the Forest Act, professional foresters found their status enhanced as well, for the first time since the heady days before World War I, when they had been looked upon as protectors of the forests. Holders of Forest Management Licences were required by law to employ foresters, and some companies had taken a renewed interest in forest management. In 1947 the Association of BC Professional Foresters was formed and, through an act of the provincial legislature, was charged with safeguarding the well-being of BC's forests. Legislation and regulations stemming from it were designed to support foresters in ensuring that forests were being well used. The integrity of the entire system of forest administration rests on the ethical behaviour of professional foresters. Their primary function is to guide the province's logging industry through an ever-shifting maze of rules and regulations, so they spend most of their working lives dealing with various state bureaucracies.

These professionals, some of them engineers engaged in the logging end of operations, others silvicultural foresters involved in establishing and tending new forests, work under a system of divided allegiances. On one hand, they have responsibility for the forests in their charge, which in theory is backed by law and their professional code of ethics. On the other hand, they must serve the interests of their employers. It is not uncommon for a forester to find him or herself torn between allegiances.

Until recently, professional associations and particularly the Association of BC Professional Foresters were dominated by an old guard that believed what was good for their employers was good for the forests. Their primary ethical concern was to avoid being apprehended while skirting rules or regulations drawn up by government bureaucrats to protect forests from overly rapacious forest companies. After opinion polls showed that the public held foresters in low regard, the ABCPF took steps to clean up its act, and a new generation of foresters began to pay more serious attention to the ethical conduct of their profession.

The other component of the forestry sector, silvicultural workers, plant 225 million trees a year, juvenile-space 40,000 or 50,000 hectares and clear brush from another 50,000–60,000 hectares. In 1998, about

18,500 seasonal workers were employed for an average of three months each at these tasks.

In spite of the disorganized state of the silvicultural sector, an enormous amount is spent on it each year. In 1992–93, the Ministry of Forests spent almost $227 million on silviculture, including about $23 million in federal funds through the FRDA II program. About 90 percent of the total was devoted to what is called "basic silviculture," the re-establishment of a free-growing young forest after logging or fire. The remainder was split evenly between activities to combat insect and disease attacks, and "incremental silviculture"—spacing and fertilizing young stands to accelerate growth rates. Those amounts declined after 1993, after the ministry underwent yet another reorganization and eliminated the Silviculture Branch. In 1996, when the FRDA II program terminated, more than $100 million in ministry silviculture costs were unloaded onto Forest Renewal BC, in spite of the fact that FRBC ostensibly was dedicated to new forest enhancement activities. As well, forest companies spend about $300 million a year replanting areas they have logged and tending those plantations to a free-growing state. Since 1987, under pressure from US lumber producers, industry has borne the costs of basic silviculture on all areas except those logged under the Small Business Enterprise Program. These costs are taken into account as part of operating costs in the stumpage appraisal process.

Until the advent of FRBC, almost all of this work was piecework, done on a contract basis. Thirty years ago, when silviculture was in its infancy in BC, the government and some of the forest companies did the job with hired crews, most of them union members. Production rates and work quality were low, and workers had little enthusiasm. As they learned they could earn more through piecework contracting, planters began to favour working in the unorganized contracting sector that appeared in the early 1970s.

Until a few years ago, BC silviculture workers were the "niggers of the woods." They were often migrant workers, BC's equivalent to California farmworkers. Neither government nor industry invested any money in their training or education. What little job security they had was threatened during every economic downturn by the establishment of relief programs that funnelled unemployed workers into silvicultural work.

Until recently, almost none of these workers were organized. After a tentative foray into the field twenty years ago, the IWA ignored tree planters. An independent organization set up in the 1970s, the Pacific Reforestation Workers Association, failed to get off the ground and eventually disappeared, and a new organization, the Silvicultural Workers Association, was formed in 1997. The major organized force within the industry is the Western Silviculture Contractors Association, representing the contracting companies. Since the 1960s a cultural gap has separated tree planters from the IWA and its members. The latter, having fought hard fifty years ago for decent living conditions for loggers and for stable employment, has been reluctant to accommodate the much different working conditions that prevail in the labour-intensive silvicultural sector. Judging by their lack of enthusiasm for the union, most silvicultural workers are more interested in the higher earnings possible through piecework and in more seasonal work patterns. For almost three decades there was a mutual suspicion between the two groups.

Silvicultural workers received few of the $15,900 a year in benefits the average forest industry worker got in 1998. Most—except for students, who are ineligible—collect unemployment insurance benefits part of the year, or they work at other jobs. Few of them are employed within commuting distance of their homes, and they rarely see the same forest twice. They travel around the province from contract to contract, living in portable camps, motels and other temporary accommodation.

The work appeals to young, healthy people with limited financial and domestic responsibilities. The nomadic lifestyle, the pleasure of working outdoors at a socially worthwhile task—restoring a forest—is attractive to some. Over time the romance fades for many, and they move on to work with a more secure future. Others, a few thousand seasonal workers, have worked as tree planters for years. Until about 1996, the annual turnover rate was between 20 and 25 percent. Beginning in the late 1980s, the silviculture industry began to mature. The planting season lengthened and some contractors took on spacing, pruning and a variety of other tasks, which allowed some workers to be employed year-round. A core of skilled, experienced workers began evolving.

Then, with the establishment of Forest Renewal BC, the stability of the silviculture sector was undermined. Ostensibly FRBC was set up to improve the productivity of BC's forests. In fact, it soon became abundantly clear that its primary function was to provide relief for loggers and millworkers laid off by other government programs, including the Timber Supply Review process, the Forest Practices Code, the Protected Areas Strategy, and so on. Several aspects of FRBC have had an adverse affect on the established silvicultural workforce.

In spite of the fact that FRBC was intended to fund incremental silviculture, the government soon transferred responsibility for $100 million worth of silvicultural programs from the Ministry of Forests to FRBC. As FRBC began to fund silvicultural contracts, experienced contract workers were laid off to provide relief in the form of training or jobs for displaced loggers and millworkers. There were now two classes of displaced silvicultural workers—union members who had lost their jobs because of other government programs, and non-union silviculture workers who were pushed aside to make way for the first group. Few of the non-union workers qualified for the relief and retraining programs set up for the union workers. Some of these programs operated out of union offices and were effectively unavailable to non-union people. The stability of the sector, hard earned by contractors and workers, was threatened. The creation in 1997 of New Forest Opportunities (NewFo), the union-run hiring hall, further undermined the stability of the silvicultural sector.

The status and working conditions of silvicultural workers are a measure of the extent to which forest management is taken seriously in BC. Reforestation appears to be viewed as a temporary obligation, best done by transient workers who can be laid off when the immediate task at hand is completed. In BC silviculture is not conducted as a business, as an enterprise devoted to the production of commercial timber or as stewardship of non-industrial forests. It is an afterthought and, on occasion, a make-work activity, a place to store the unemployed so they do not put up too much of a fuss.

Using silvicultural funds for unemployment relief during the recession of the early 1980s was counterproductive. Most workers were not highly motivated to undertake the arduous labour involved in tree planting, spacing and pruning, and most of them drifted off into other occupations after the funds were used up, rather than finding employment with established contractors. The same pattern could be observed during the early stages of FRBC. While some displaced loggers and millworkers were eager to learn silvicultural skills, just as many were not. They were too old, too out of shape or too uninterested to put in more than the minimal effort required to collect their relief cheques. When the "retraining" projects ended, they moved on to other work, leaving behind some marginally acceptable, but quite expensive, work.

Perhaps the greatest failure of BC silviculture efforts is that after spending several billion dollars over the past twenty-five or thirty years, the province still does not have the silvicultural capability needed to tend the forests well enough to sustain or enhance the forest economy. And FRBC, an agency ostensibly established to improve forest stewardship in BC, has instead become, at best, a waste of money and, at worst, another obstacle to such stewardship.

Other Players

In addition to the major forest industries, there are various other industrial and non-industrial users of the forest. The forest products industry includes producers such as pole and piling manufacturers, log-building makers and shake and shingle operators. The latter, for example, exported $243 million worth of roofing materials in 1993, almost all of it to the US.

In recent years a brisk commercial trade has developed in non-timber, botanical forest products, including mushrooms, floral greenery, medicinal plants, berries, herbs, landscape products and craft items. Although it is difficult to assign a monetary value to these products, the Ministry of Forests estimates the annual mushroom harvest alone generates revenues of about $20 million.

Beyond this, the values we obtain from forests are difficult to quantify. Various attempts have been made to estimate the dollar value of forest-dependent recreation, tourism, water resources, fisheries, wildlife and so on. Because there is no mechanism for charging for these values, there is no measure of their worth comparable to that for timber products. This is not to say they are worthless, only that it is difficult to include them in economic measurements—which may say more about the economic methods used than the status of non-timber forest products.

This is the shape and dimension of the BC forest economy. In addition to employing a sizeable portion of the provincial workforce at above-average rates of pay, it pumps substantial amounts into the coffers of various governments each year—$3.7 billion in 1998 (see Table 4B).

The forest industry is, clearly, not a sunset industry. But is it becoming one? In order to sustain itself, any industry needs to maintain sufficient capital investment to replace worn-out and outmoded production facilities. Some analysts believe the BC forest industry requires about $3 billion a year in capital investment to keep itself healthy. Mills must be rebuilt, new technology must be installed, pollution control systems must be put in place, workers must be provided with the tools they need to compete in the marketplace. After 1990, when capital costs of $2.25 billion were incurred, investment in the forest sector declined dramatically; in 1998 it stood at only $729 million. This level of capital spending means the industry is currently living off its capital, a situation that will compromise severely its economic future as mills and logging equipment wear out.

The forest industry cannot continue to function as it has in the past. In partial recognition of this, the Ministry of Forests has reduced the maximum amount of timber harvested in some supply areas each year. Also, in view of the impact of conventional logging techniques on forest ecology, substantial areas of forest land have been removed from the commercial forest land base in some supply areas and designated as parks or given other protected status.

These changes by themselves will reduce the scope and scale of the industry. And that is one possible option for the future—that BC simply forgo some of the economic benefits derived from forests in the past. We can adjust to an annual timber harvest reduced by 20 or 25 percent, if we accept that the provincial economy will also be reduced by 20 or 25 percent. Forest-based jobs will decrease proportionally, as will public revenues and the foreign exchange earned from exports. Displaced workers will find other jobs, likely at lower wages. Public sector and many private sector wages will be reduced proportionally, and the 75 or 80 percent who keep their jobs will count themselves lucky.

Certainly, some jobs can be created in the value-added sector. The legislature's Select Standing Committee on Forests issued a report in 1993 stating that with substantial restructuring of the industry, employment in the value-added sector could be doubled. So far, however, after the expenditure of vast sums of money through FRBC, employment in value-added companies remains constant at about 12,500. And in any case, recent enthusiasm for expansion of the value-added sector, in part fuelled by the availability of money from FRBC to develop value-added plants, has acquired some of the characteristics of religious revivalism instead of the hard-headed analysis required.

There are other options. BC can use its commercial forest lands as others in the world do, putting more people to work in the forests growing more and better timber, which can be used in mills to manufacture even more valuable products. And forest values can be assured at the same time.

These options are available, but as a closer examination of BC forest politics reveals, developing them will not be easy.

5 Public Participation

Nothing stirs the passions of British Columbians more than their forests, and since the late 1960s they have been increasingly insistent on participating in making decisions about those forests. In no other sector of public administration is the desire for involvement so pronounced. But in spite of this greater interest and three decades of intense lobbying, provincial bureaucrats have managed to keep the public a long arm's length away from decision-making.

Not long ago it was widely held that in our system of representative government, once we had elected our politicians they governed as they saw fit and faced the consequences of their actions at the next election. They set policies and made difficult political decisions, then oversaw their implementation and administration by civil servants.

But by the 1960s, the country had become so large—with an even larger government—that people wanted to play a more active role in decision-making processes that profoundly affected their lives. This was not just a BC phenomenon; it was going on in all Western democracies and could be felt all over the world, as could the resulting conflict between citizens and governments.

The arena of forest resource planning in BC has been no exception, and British Columbians are still testing democratic institutions and definitions of public participation. Citizens first rejected the idea that the future of the forests could be left in the hands of elected representatives and their hired experts. Then other possibilities emerged. One was that people needed only to be informed about decisions being made on their behalf, while they were being made. This did not long satisfy the need for democratic process. Another method was to allow people not immediately involved in forest development to express their opinions, but still to leave the decision-making to the bureaucrats.

A third approach, and a controversial one, is to relocate the act of making decisions closer to the citizens it affects. This devolution of power has rarely if ever been tried in BC, and it is a threatening idea for many of those in charge. Some bureaucrats and technocrats resist interference in their domain. Many people believe important policy decisions should be made in Victoria to achieve consistency throughout the province, and that all important resource management decisions should be made by skilled, knowledgeable people acting in the best interests of the BC public. The attitude persists that if decision-making processes were decentralized to local bodies, these bodies would not exercise the same public-spirited foresight as experts answerable to Victoria; they would intentionally or unintentionally plunder the resource base and move elsewhere.

The belief that ordinary people cannot be trusted and must be saved from their own greed and stupidity may be a particularly Canadian one, and a reminder that we are still colonials at heart. We have a habit of placing our trust in a wise and benevolent authority, which presides over us. The monarchy assumed this authority before democratic governments evolved. Today more and more Canadians, including British Columbians, believe that monarchs and central governments cannot deliver on all of the responsibilities they have assumed, and that a more direct form of democracy is desirable.

What is meant by the phrase "the public"? In general usage it is not to be equated with the citizens of a country or province, nor with any individual person. Instead it has become a convenient abstraction, as in "the public interest," which does not mean any particular group's interest but refers to some abstract interest. "The public" does not, in fact, exist. In BC, when we talk of "public land" or "publicly owned forests," we are not talking about land that you, I and our fellow citizens own, we are talking about land owned by that amorphous entity the state—or, as we say in Canada, the Crown. Very often when people talk about the "public interest," they are in fact talking about the interest of the state.

It is important to distinguish between the members of the society, the government they elect to represent them, and the administrative apparatus—the bureaucracy—established to manage government affairs. The interests of the people are often confused with the interests of "the government," which is actually a group of people belonging to the same political party, much of whose interest is focussed on retaining their position of power. While it is true that most politicians are genuinely concerned about representing their constituents, it is a simple fact of political life in the late 1990s that the dominant factor in their day-to-day activities is the need to act in concert in order to maintain their position as "the government." By extension, the government of the day is often not distinguished conceptually from the relatively permanent administrative apparatus that has evolved over time to manage the entity known as "the state." It is even assumed sometimes that a bureaucracy's interests are identical with those of the citizens. In fact, bureaucracies tend to look upon individuals or groups of citizens not as part of "the public" whose interest they exist to serve, but as separate, private entities whose interests are distinct from and possibly at odds with "the public interest." Former NDP Forest Minister Bob Williams recently commented on the growing tendency of senior civil servants—sometimes referred to as "the Mandarins"—to view themselves as the ultimate representatives of the "public interest."

All of this confusion has contributed to the evolution of state administrative organizations—bureaucracies—as bodies that tend to behave as though they are corporate entities in their own right, with interests and agendas of their own. Thus, the Ministry of Forests at various times looks upon the government as its enemy, as Chief Forester C.D. Orchard did in 1952, when the first Social Credit government was elected: "In my opinion the Socred government was a one man administration. Ministers were nonentities who referred the simplest routine matters to Mr. Bennett, the Premier, and parroted his decisions. Certainly the new government took office firmly of the opinion that every predecessor in office had been venal and that no civil servant could be trusted."[1]

In a similar manner, the Forest Service came to look upon the citizens of BC, if they were not part of the forest industry, as interlopers in the forest sector. By the 1960s, when increasing numbers of citizens began appearing at Forest Service offices asking for information, the bureaucracy was clearly operating on the assumption that the province's forests were the property of the state, or the Crown, and as such they "belonged to" the Forest Service. Hence, a Forest Service manager's answer to my request in the early 1980s for a Woodlot Licence on Crown forest land: "No. These are our forests. We will manage them. The private sector has Tree Farm Licences. These are public forests. They are ours."

Prior to about 1970, forest resource management decisions were made behind closed doors by the Forest Service and the forest industry, with the occasional participation of elected politicians. The doors were not closed to keep the public out: only rarely did they want in. From 1912 until the 1960s, the administrative apparatus in the Forest Service was relatively decentralized, with each Ranger District looked after by a forest ranger, who was accountable for protection, logging and log-scaling practices, collection of revenues, and everything else concerning the forests in his district. Rangers spent a good deal of time in the field and were close to the forest and the people who lived and worked there. Anyone wanting to know what was going on in the district's forests could corral the ranger and ask him.

But with the adoption of the 1948 Forest Act and the rise of large, integrated forest companies, the authority and effectiveness of rangers began to erode. Their on-site decisions, which had once been final, were routinely countermanded. In a typical scenario, a company's chief forester would take a ferry over to Victoria, have lunch with a senior member of the Forest Service and get the matter resolved to his satisfaction.

A perfect example of this process occurred in the early 1980s, even though by that point the procedures were being changed and the Forest Service reorganized. A conflict arose between the citizens of Kyuquot, a small community on the west coast of Vancouver Island, and several large forest companies with cutting rights in the nearby Tashish River watershed. In one of its earliest attempts at involving a local community in the planning process, the Ministry of Forests' district office in Campbell River set up a task force composed of two Kyuquot residents, representatives of each forest company and a half dozen government departments, and three or four environmental organizations. Their objective was to draw up a land use plan for the valley to send to the provincial cabinet. Because there were about 30,000 cubic metres of unallocated cutting rights in the Timber Supply Area where the Tashish was located, land could be taken out of the working forest without reducing any company's cutting level. There was great potential for an easy solution.

Few of the participants had had any experience with this kind of process, so it was long and arduous. After almost two years, during which fifteen to twenty people sat through twenty-five or thirty day-long meetings, they began to feel they were reaching an agreement. At that point, unknown to staff in the Forest Service's district office, a deal was cut in Victoria. A decision had been made earlier to reduce by 30,000 cubic metres Crown Zellerbach's allowable annual cut in the Queen Charlotte Islands. CZ was also a participant in the Tashish task force. At a discreet private luncheon meeting in Victoria, CZ's chief forester and an assistant deputy minister of forests agreed to replace CZ's lost cutting rights in the Queen Charlotte Islands with the 30,000 cubic metres available in the TSA where the Tashish lies. The possibility of adopting an agreement worked out by the local participants was eliminated by the upper levels of the same bureaucracy that had established the task force. Ironically, a large part of the Tashish was made into a park ten years later and the timber supply reduced accordingly.

By the time Peter Pearse held his Royal Commission hearings in 1975, it was clear that a lot of people outside the forest industry wanted to be involved in the forest planning process. The term "planning process" was a euphemism for the decision-making process: it sounded better to say that a group of technocrats were sitting around a table drawing up a plan than to suggest they were deciding what areas would be logged and by whom. It is still useful, twenty-five years later, simply to substitute the word "deciding" for the word "planning" when trying to understand what is happening in the forest sector.

Pearse received many submissions from non-timber interests concerning the use of the forests, and many from outside the industry followed his commission closely. Until about that time, in fact, one of the chief complaints of the industry had been the public's lack of interest in forests and forestry, allowing politicians to ignore forest issues and funding. Before long that complaint would disappear, to be replaced by a fervent desire for a renewed public indifference.

Pearse recognized the problem as "not so much the need to expand the total effort in terms of coverage and time, but rather to find systematic and practicable methods of integrating with traditional forestry planning the necessary provisions to protect and enhance forest values other than timber."[2] He recommended the establishment of locally based representative bodies to assess resource plans, and for Regional District Boards to set up resource advisory committees that would monitor the activities of resource bureaucrats[3]—in other words, for a devolution of power to local government. His advice was never followed.

In the wake of Pearse's report, a group of Slocan Valley residents made one of the first attempts in BC to establish a formal planning review process by a local citizens group. The proposal got lukewarm support from the Dave Barrett NDP government, but it was stonewalled to oblivion by resource sector bureaucrats, particularly those from the Ministry of Forests, who wanted no truck with real citizen involvement in their bailiwick. By that time the ministry had begun, reluctantly, to discuss its plans with other government agencies, but it was not about to do so with taxpaying citizens.

The drastically revised 1978 Forest Act opened the closed doors of the planning process a crack. Even so, the only provision for wider participation defined in the new act required the Forest Service to hold public hearings before granting new Tree Farm Licences or Pulpwood Agreements, and the intent of this clause was to accommodate other interested parties from the forest industry, not concerned citizens or non-timber interests.

The Ministry of Forests Act, passed at the same time, contains a remarkable example of the attitude of contemporary government agencies toward the citizens they are supposed to serve. The ministry is instructed to perform its functions "in consultation and cooperation with other ministries and agencies of the Crown and with the private sector."[4] By way of explanation, then-Deputy Minister Mike Apsey, who oversaw the framing and implementing of the Act, explained that the "private sector" includes the public. In other words, the Ministry of Forests and other government agencies sit on one side of the table, and private interests, including that vague entity "the public," sit on the other side. If ever anyone from "the public" approaches the Ministry of Forests, they are, ipso facto, acting in a "private" capacity and are dealt with accordingly by the ministry, which in theory is caring for forests owned by "the public." No wonder the Ministry of Forests has, for twenty-five years, behaved as though it owns the forests and the citizens have no business participating in making decisions about them.

By the early 1980s, an increasingly vocal and organized citizenry was battering at the doors, which were now firmly closed. Bowing slightly to the pressure, the ministry hired a "public involvement coordinator" in 1980, and the following year published a handbook describing how the formal planning process worked and where and how ordinary citizens could participate. Apart from special situations, where task forces or study groups would be set up to examine specific resource use conflicts, the opportunities were very few. And naturally there was no mention of the kind of informal processes by which 30,000 cubic metres of Tashish timber had been allocated to Crown Zellerbach at about the same time the handbook was prepared.

At the provincial level, under Bill Bennett's Social Credit government, everything was done behind closed doors. The main planning instrument at this level was the publication of a "Five-year Forest and Range Program," drawn up within the confines of the ministry to design policies and programs for provincial and regional managers. Establishing allowable annual cuts in Timber Supply Areas and Tree Farm Licences was part of this program, a task performed in splendid isolation by the chief forester.

At the regional level, provincial plans and programs were applied in conjunction with industry and other government agencies, with no provision for the involvement of others. By this point in the process, broad land use policies had been established, the amount of logging for each yield unit had been determined, and numerous other important decisions had been made, many of them with the active participation of forest companies and bureaucrats from other government departments.

Finally, at the TFL and TSA planning level, once every five years "the public" was invited in for a few carefully controlled glimpses at what was in store for the province's forests during the next half decade. Public input was requested at the beginning to identify resource use issues and management objectives, and certain documents were available for examination, although most of the information upon which the decisions were based—such as inventory data—was not normally available. Once a draft of the area's five-year plan was drawn up, the public was allowed to read it and make written submissions, which were forwarded to the chief forester when the plan was sent in for his approval. Typically, few outsiders bothered to examine these plans; this was invoked as an argument against the need for public involvement, on the grounds there was no interest.

If the plan was for a Tree Farm Licence, that was the end of public participation. Further involvement was allowed at the local resource use planning level in the Timber Supply Areas, where plans for smaller areas such as watersheds were determined. The type and degree of involvement was determined by the amount of conflict the situation might generate. For example, in the Tashish, when local citizens objecting to the plan began to gather support outside their community, a task force was established. Prior to that, and wherever this degree of citizen activity was absent, there was no means of taking part in the process. Hearings and task forces could be part of the process, although they were limited to providing the ministry with advice. After this stage, intergovernmental planning teams took over to work out their various concerns, a process from which ordinary people were excluded.

At the final, operational stage, when cutting plans for specific logging sites were drawn up, the completed plan had to be displayed to the public and to other government departments. The Forest Service district manager was required to consider comments on the cutting plans but was not required to accommodate them.

That was the sum total of public involvement in determining the use of BC forests until the early 1990s. The Ministry of Forests' position always appeared to be that public—that is, private, non-governmental—interests, as defined in the Ministry of Forests Act, would eventually lose interest in forests and go away. Then the decision-making process could return to its old format. Non-timber interests and values could best be taken care of inside the bureaucracy, the ministry felt, by co-ordination and co-operation with other government agencies. If people were worried about fish, the federal fisheries department would take care of their concerns; if they were concerned about the environmental impact of logging, the Ministry of Environment had the matter well in hand. These agencies, after all, were acting in "the public interest." In the 1984 *Forest Range & Resource Analysis* report, written amid several heated battles over forest use, the section on conflicts between timber and non-timber interests refers only to interactions with other government bodies. There is no mention of citizen participation in the entire 300-page report.

Even though opportunities for public involvement expanded in the 1980s, the intransigence of the ministry and the forest industry fuelled a growing frustration. They simply proceeded as before, in a process that became known as "talk and log," which meant allowing people to participate in a limited way while business continued as usual.

Therefore, those opposed to logging in watersheds or other logging practices found the only avenue open to them was to mobilize media attention and encourage voters to bring pressure to bear on the politicians, who would eventually order the Ministry of Forests to modify its plans. Usually this meant withdrawing the area from commercial use, either temporarily, while a "study" was undertaken, or permanently. Diehards in industry and the Forest Service were astounded to find areas they had considered secure for logging, such as the Carmanah and Stein valleys, were now declared off limits. Both sides dug in their heels, and by 1990 a public relations battle of huge proportions was being fought over several proposed logging operations, particularly those planned for pristine, old-growth watersheds.

At the same time, the forest industry was beginning to find the official public involvement process equally dissatisfying. The process could not accommodate public expectations, so it often dragged on endlessly with no resolution of conflicts. Major differences were ultimately referred to elected officials, where the courage to make decisions was often lacking as politicians were forced to choose between two or more opposing sets of voters. Participation in the process and operational delays were costly. Many in industry felt that genuine citizen involvement in the planning process, including the participation of people who worked in the forest industry, was preferable to the increasingly bureaucratic process then in place. At this stage the Bill Vander Zalm government was still in power, but in such disarray it was in no position to cope with mounting resource use conflicts.

The Forest Resources Commission, in its 1991 report, dealt extensively with the planning process and the public's role in that process. Consistent with its propensity for bureaucratically expansive solutions, the FRC recommended that a new Land Use Commission, including a hierarchy of regional and local planning groups, be established to draw up and administer a province-wide land use plan. If adopted, it would have created a highly centralized planning system, with clearly defined procedures for input from local and regional planning agencies, as well as legally defined procedures for public involvement at local and regional levels. It would have ended the Ministry of

Forests' control of the forest planning process, relegating it to an advisory role on a par with other government departments. Like the rest of the FRC report, this recommendation was shelved.

The Harcourt government tackled the situation head-on. Its response to this issue, which had peaked at about the same time the new administration took office in 1991, was to establish the Commission on Resources and the Environment (CORE) with then former provincial ombudsman, Stephen Owen, as commissioner. It was set up in mid-1992 as a quasi-independent body that reported to the legislature and had authority to issue reports on a wide range of land use issues. It was given a powerful role with a broad mandate to develop a provincial land use strategy, and to establish regional planning processes with a high level of public participation. Essentially, CORE took over major components of the provincial and regional land use planning process previously handled by the Ministry of Forests and other resource agencies. Empowered to report directly to the legislature and the public, it was intended to integrate the management of the province's resources from a position of authority.

During its first three years of operation, CORE worked on three distinct levels. First, its staff drew up a series of papers and documents defining a provincial land use strategy. Second, it organized four regional "tables"—Vancouver Island, Cariboo–Chilcotin, East Kootenay, West Kootenay–Boundary—to develop regional land use plans with the active participation of representatives of interest groups. And third, it helped organize smaller community groups to develop local plans within the context of the regional plans.

For two years many people were involved in the CORE process. Never before in the history of BC had so many people representing so many varied interests participated so intensely in planning anything. If it accomplished nothing else, CORE temporarily ended the conflict between environmentalists and the forest industry simply because almost all the players were so caught up in the process they had no time for anything else.

The regional tables were the scene of most of the visible activity. "Seats" at these tables accommodated a wide range of interests, which varied in each of the four regions. At the Vancouver Island table, for instance, seats were provided for agriculture, conservation, fisheries, forest unions, other unions, independent forest interests, major forest companies, local governments, mining, provincial government departments, recreation organizations, economic development groups, tourism and youth. Each of these sectors brought to the table four or five spokespersons and a dozen or more members of a steering committee—200 to 300 people for each table. They met monthly for up to two years in marathon sessions of three or four days each, twelve to fourteen hours a day.

For the most part, CORE was a sincere attempt by representatives of various interest groups to arrive at consensus on how land in each of these regions should be used, but instead of being asked to define and articulate values and objectives, the participants were asked to make detailed decisions about which resources in which areas to use for what. Support facilities required by the consensus process were not available. And too many of the participants had no real commitment to the process. At the Vancouver Island table, for instance, some participants, such as spokespersons for environmental organizations that did not themselves use forest resources, had nothing to lose from the process and everything to gain, while for most forest industry representatives, the opposite was the case. At other tables, regional spokespersons found themselves caught between local and provincial interests. In the West Kootenays, whenever local representatives of the major forest union, the IWA, began to reach a working compromise with local non-timber forest users, spokesmen from the union's head office in Vancouver would intervene and reassert the union's intransigent stance, which was to oppose the removal of any land from the commercial forest base.

Also undermining the entire effort was the fact that government had not devolved any decision-making authority to these tables. CORE was required only to submit to government a report of the deliberations. Each table was promised that if it could reach consensus, the government would adopt its recommendations. If they did not reach consensus, the CORE commissioner would produce recommendations anyway, which the government might adopt. Some participants with good access to the government went through the entire process knowing that no matter what might be agreed to at the table or put in the commissioner's report, they had further opportunities to influence the government's final decision. Other interests, such as some environmental organizations, declined even to participate in the exercise for fear of their positions being compromised, leaving them free to continue their campaigns after the process was finished, regardless of the outcome.

In the end, none of the CORE regional tables reached consensus, although some came close. The commissioner submitted his reports to the government, which quickly produced a land use plan for each region. These plans were announced by Premier Mike Harcourt with great fanfare and ceremony. They incorporated some of the agreements made at the CORE tables, decisions on new parks arrived at under the concurrent Protected Areas Strategy, and other matters from government forest sector initiatives such as the Forest Practices Code, Forest Renewal BC and the Forest Land Reserve.

To complicate matters, by the time the CORE regional process had wrapped up and the government produced its plans, the forest policy climate had evolved and, in some cases, moved beyond the policy issues dealt with by the CORE process. Adoption of the Forest Practices Code, for example, affected cut levels in the regions, in some cases undermining whatever consensus had been reached at the tables. More important, the government acknowledged that its land use plans were subject to revision as aboriginal land claims negotiations proceeded. In some regions, Natives had declined to participate in the CORE process because they wanted to negotiate their claims directly with the government, not at a table with other claimants to the land.

The plans announced by the government were not actually plans at all. At best they were vaguely defined outlines of intent. For instance, the Vancouver Island "plan" said that within the 81 percent of the island dedicated to commercial forestry and other resource use, there would be separate zones of low-intensity and high-intensity forest management. But the plan did not identify these areas or define the levels of management. A committee of bureaucrats from the ministries of forests, parks and environment was set up to fill in these details by the end of 1994. By mid-1999, large sections of the Vancouver Island land use plan had still not been implemented.

At the same time it was operating its regional tables, CORE was also organizing at the subregional or community level. Its major function here was to encourage the formation of "community resource boards," local bodies consisting of volunteers from a wide variety of special interest groups, industrial users, local governments, Native groups and senior government agencies.

CORE initiated pilot projects in a number of BC communities to find out whether local groups could draw up local resource management plans within the broader context of the regional plans being negotiated at the regional tables. At Anahim, in the Cariboo–Chilcotin region, twenty-nine local interest groups, including local governments, tribal and band councils, recreation associations, forest companies and unions, community resource boards, cattlemen, trappers, guides and outfitters, various provincial government departments including the Ministry of Forests, the federal Department of Fisheries and Oceans, and numerous other community groups, spent a total of fifty-two days in meetings and came up with a plan to which all agreed. It was incorporated into the regional table's agreement, which went on to CORE and, eventually, into the regional land use plan adopted by the government. Another team at Chilko Lake spent two years reaching consensus to establish a park and defined resource management area, which the government did in 1994.

A similar pilot program was set up by CORE in the Slocan Valley, in the West Kootenay–Boundary region. The composition of this group was quite different from that at Anahim, and did not include government representatives, although the provincial forests and environment ministries did provide technical support. The Slocan Valley community was much more polarized over forest management issues and failed to reach consensus on a plan within the time frame established by CORE. Considering the long-standing animosities that divide this community, perhaps more so than any other in BC, this failure is not surprising.

The Vancouver Island plan included a decision by the government to establish community resource boards composed of volunteers from the private sector, local government and Native bands to give advice on implementing the regional plan.

While hundreds of people were engaged at the regional and local round tables, the CORE staff were busy producing a flood of documents outlining a provincial land-use planning and administrative structure. The best known of these efforts was probably the idealistically worded Land Use Charter, which the government was persuaded to adopt in principle. This charter was a proposed organizational structure to co-ordinate and integrate certain functions of other government ministries and agencies. As one CORE document stated: "This can best be managed by an agency solely responsible for the coordination and integration of land and resource management ministries. This agency could take the form of a secretariat headed by a deputy minister and reporting directly to cabinet; a separate ministry; or an independent commission."[5]

This secretariat would have functioned as a parallel bureaucracy, devoted to the task of inter-bureaucratic co-ordination and, of course, direction. It would report directly to cabinet. In concept, it was similar to the provincial cabinet's Environment and Land Use Committee (ELUC) secretariat, which had flourished briefly under the previous NDP administration of Dave Barrett. The proposal for the new secretariat was not a surprise, as CORE's first deputy commissioner was Denis O'Gorman, the former head of the ELUC secretariat. CORE's modestly defined role for itself in this brave new bureaucratic world was that of providing "independent oversight of the fairness and effectiveness of government administration."[6]

By late 1994 or early 1995, when these documents were completed and the government had drawn up its various land use plans, CORE had fallen out of favour in Victoria. It was seen as an expensive fomenter of conflict over land-use issues. Perhaps most telling, the established bureaucracy was able to convince government there were better ways to make plans and keep the "private" sector—i.e., the public—at bay. During 1996, after Glen Clark became premier, the rug was deftly pulled out from under CORE. Commissioner Owen leapt sideways to another senior bureaucratic position and by the end of the year the staff was laid off and the office closed. Like the "permanent" Forest Resources Commission, it quietly faded out of existence. The millions of taxpayer, corporate and private dollars spent on its operations had produced no lasting accomplishment. Those close to the process agree that CORE did bring together a lot of people who previously had been in conflict. After CORE, they at least recognized each other when they met on the street. To some extent the polarization of BC communities over resource use disagreements had been arrested, but the land use planning process had not changed substantially.

On reflection, CORE's fate was probably sealed by the parallel growth of another planning process led by the Ministry of Forests. While CORE was setting up its round tables, an inter-ministerial planning group was organizing Land and Resource Management Planning (LRMP) groups in twelve Interior Timber Supply Areas. These groups also involved representatives of a range of "private" interests and government agencies in a consensus-based process to draw up resource use plans for Crown lands. The intent was to put in place by the year 2000 LRMP groups for those areas of the province not covered by CORE-built land use plans. This process was supported by another Victoria agency the government established early in its mandate, the Land Use Co-ordination Office (LUCO). It was staffed with senior bureaucrats directed to co-ordinate the large number of inter-ministry programs being devised by the NDP government and its rapidly expanding army of civil servants. In spite of the demise of CORE, the forest planning process was still alive and, once again, firmly in the hands of the entrenched bureaucracy.

Underlying this process, and at times circumventing it, was another method developed by the Clark government, one that could be called the "stakeholder system." It consists of inviting select interest groups to a series of small private meetings with senior bureaucrats and politicians, and eliciting their opinions and reactions to initiatives or programs the government is contemplating. The stakeholders are not brought together to discuss the issues, nor are public meetings held, at least until real decisions are made. Stakeholders are consulted as plans are refined, primarily to insure their acquiescence to developments. Nothing is revealed until key stakeholders have signed on to a proposal. This was the process used in the early months of 1999 to consider a number of major forest policy revisions, including changes in the tenure system.

Perhaps what the CORE saga revealed was the difficulty, if not impossibility, of running a geographical entity as large and diverse as BC, in a manner that is responsive to citizens and taxpayers, through a centralized state bureaucracy. Almost the entire provincial land base is owned by one level of government. As this land has been developed over the past century, its owner has been forced to cope with the growing number of problems associated with managing it. This has led to the mushrooming of large provincial bureaucracies, such as the Ministry of Forests with its 4,500 employees. If this were a private corporation, under the direction of a capable, competent board of directors, the task would be daunting. Because it is under the direction of governments, which are controlled by political parties, whose primary concern is re-election every four years or so, the undertaking is impossible. Conflicts and disagreements develop; the bureaucracy defends what it sees as its turf and finds itself at odds with the citizens it theoretically exists to serve. Citizens turn to their political representatives; bureaucrats are ordered to proceed in yet another direction. By the time they figure out how to circumvent government red tape, or to deal with the new demand in a manner that does not conflict with existing laws, regulations, budgets or province-wide directives, some new conflict has emerged and the cycle begins again. The result is frustration, ill will and a growing reluctance to make decisions, which threatens the foundations of the provincial economy.

Part of the solution to these frustrating, confounding problems lies in some authentic delegation or devolution of authority for resource planning to municipal and regional governments that have been elected by the people living and working in the communities where the resources are found.

6 Tenure

Among the most important aspects of British Columbia forest policy are those concerning tenure, or access to timber and forest land. Laws and regulations dealing with tenure have dominated the agenda at the last four public inquiries into provincial forest policy, and commissioners have identified tenure as one of the most important issues. In the late 1990s the lack of a workable tenure policy remained as one of the great shortcomings of BC forest policy.

The objectives of forest tenure policy have changed over time, with one dramatic turning point. Prior to 1947 provincial tenure provisions were designed to supply industrial timber to the forest industry and to raise revenues for the government. The benefits of this policy were the economic development of a young province, including employment and business profits, and the creation of public sector infrastructure. First-phase tenure policy did not address itself to the fate of either the existing forests or those that might replace them. In some cases, where non-timber interests were strong and little commercial timber was growing, parks were created. But conservationist principles had little effect on tenure provisions in BC. It was understood and accepted that forests, particularly those in the most accessible and productive locations, were being liquidated to make way for agriculture.

After 1947 a much different principle underlay tenure policies: sustained yield forestry, with the objective of maintaining a steady supply of timber. New forms of tenure were needed to encourage forest management and the production of new commercial timber crops.

Before BC joined Confederation in 1871, the only way colonial administrators could transfer ownership of timber from the Crown was to grant title to the land on which it grew. The first sawmills at Alberni and Burrard inlets were provided with timber in this manner. The same method was used after Confederation to facilitate railway construction, when more than 6.5 million hectares of mainly forested land were granted to the federal government. This was some of the most productive forest land in the province, and because it was close to rail lines, the timber on it was readily accessible. The most valuable timber stands were sold to forest companies; in fact, much of the forest industry was established on this timber. After the land was logged, much of it was allowed to revert to the Crown in lieu of taxes. Today, only about 2.6 million hectares of these railway grants remain in private ownership. A third of this private land is forested and the remainder has been used for settlement or agriculture.

Gradually, near the end of the nineteenth century, the government cut off the sale of timber land. In 1887, anyone purchasing land had to declare it was "not chiefly valuable for timber." Of course, many false declarations were made and the sale of timber land continued. By 1896 the sale of timber land, defined as land containing 8,000 board feet of timber per acre on the coast and 5,000 board feet in the Interior, was prohibited outright. This provision was not rigidly upheld until 1912 when the Forest Service was formed and the law enforced. Since that time, there has been no means of purchasing or otherwise acquiring outright ownership of Crown forest land in BC. But there has been a substantial transfer of forest land back into Crown ownership during that same period, some because of non-payment of taxes, and some by government purchase.

Prohibitions on land sales did not affect the forest industry adversely and were not intended to do so: other means of obtaining timber had already been devised. As early as 1865, in the colony of Vancouver Island, the governor passed a land ordinance permitting colonial administrators to issue Timber Leases, which granted the right to cut timber on designated areas, with title to the land remaining with the Crown. An annual rent was charged for the leases, which could be of any size, and they could not be sold or otherwise disposed of. This tenure instrument was used by the united colonies of Vancouver Island and mainland British Columbia, and then by the province after 1871. Between then and 1905, four other forms of timber-granting tenure were devised.

Timber Licences, to a maximum of 400 hectares each, were introduced in 1888 and their terms modified at various times after that. The licensee paid a modest annual rental and a royalty for the timber after it was harvested. In 1905 the McBride government allowed Timber Licences to be sold, initiating a speculative timber staking frenzy. Aware that timber was running out in eastern North America, investors snapped up the leases in anticipation of a growing forest industry in BC. Although it is tempting now to be critical of the large profits some of these speculators earned, their involvement did generate value for forests that had previously been considered worthless. Allocation of the Special Timber Licences, as they were called, was terminated two years later after 15,000 of them had been issued.

Around the turn of the century the government allowed some Timber Licences to be converted into Pulp Licences, charging lower royalty payments on low-grade pulp wood. Between 1901 and 1903, several very large Pulp Leases were also granted to encourage the establishment of a pulp and paper industry in BC, although it was not until 1909 that the first pulp mill opened at Swanson Bay.

In 1886 the government introduced the handlogger's licence, allowing an individual to cut timber from the shoreline for one year, with payment of an annual fee plus a royalty. Until 1907, handloggers were free to cut on any unallocated land, the way a hunting licence allows hunting anywhere. After the first few years, handlogging was more strictly defined: no power equipment was permitted, and the technique was to select trees growing on steep slopes and fall them so they would drop or slide into the water, sometimes with the help of hand jacks. In the pioneer era of coastal logging there was a large contingent of colourful handloggers. In 1917, 388 of them working along the entire coastline selling logs to independent mills accounted for 2 to 3 percent of the provincial timber harvest. In modified form, handlogging licences have survived until the present day, but they constitute such a small part of the industry they are considered insignificant in discussions of tenure policy. They were important in the early years of this century, however, because they provided entry into the industry for entrepreneurial loggers, many of whom became successful on a larger scale.

By 1907, about 4 million hectares of forest land was "alienated," as bureaucrats describe land held by private citizens. Most of it, about 90 percent, was in the form of Timber Licences. The four types of tenure, long referred to as "old temporary tenures" because they expired when the timber was removed, were a form of area-based tenure: they allowed all the timber on a specified tract of land to be harvested. In a similar manner, the federal government retained ownership of its land in the Interior railway belt and allocated logging rights in the form of Timber Berths. When the railway belt was returned to the province in 1930, BC assumed title to the land and administration of the timber tenures.

In 1907, the McBride government realized the 4 million hectares of forest land then available to the embryonic forest industry in the form of Crown grants or temporary tenures would supply it for the foreseeable future. The best accessible timber was already claimed, and further allocations would only devalue timber already available, driving royalties down. The government suspended further allocations of timber tenures, and until 1912 there was no means of "alienating" Crown timber. That year the Fulton commission was appointed to study the tenure question, along with other forest policy issues.

Fulton estimated the industry then controlled about 160 billion board feet of timber and used only 1 billion feet a year. He recommended the remaining unallocated timber be kept in reserve, except for small parcels to be sold competitively at sealed tender or public auction. Known as Timber Sales, these tenures were originally intended to play a minor role in timber allocation, but gradually assumed a more important function and by 1945 accounted for 25 percent of the timber harvest. Pulpwood Timber Sales, a little-used class of Timber Sale, were also permitted. They were, in fact, grants of timber to pulp mills at reduced royalties, and by 1945 more than 30,000 hectares were covered by pulpwood sales.

These tenures, consisting of land grants, old temporary tenures and Timber Sales, were industry's source of timber until the Forest Act was rewritten in 1947 in the wake of the first Sloan Commission. Because all types except Timber Sales were permanent or renewable, some of each type survived until that time, and beyond.

The fundamental question of land ownership did not arise as an important issue until 1947, when proposed revisions to the Forest Act were circulated. Until then, the forest industry's only need was for mature, commercial timber to keep its mills supplied, and there was plenty available from speculators and the Esquimalt & Nanaimo Railway Company. From the late nineteenth until the mid-twentieth century, the policy of government ownership became more firmly entrenched.

Somehow Canadians ended up with most of their forest land state owned, unlike most other countries in the world. When the policy was enacted, there was no evidence anywhere that state ownership produced superior levels of forest management or stewardship. If anything, the record showed private forests were better managed than those held collectively. "The tragedy of the commons" is a phrase widely used to describe the fate of state- or community-owned land. A century's adherence to a policy of state ownership in Canada suggests this policy is more of a hindrance than an aid to good forest management. How, then, did the policy come to be adopted throughout the country, and why is it still so vigorously defended?

GLOBAL FOREST LAND OWNERSHIP
(By percent)

Country	State	Private	Corporate	Other
Norway	18	75	—	6
Finland	24	65	7	4
Sweden	26	49	25	—
France	12	70	—	18
U.K.	50	50	—	—
W. Germany	31	44	25	—
Yugoslavia	40	60	—	—
Former USSR	90	—	—	10
Japan	32	57	—	11
US	28	59	13	—
Canada	92	(8	)	—
BC	96	1	3	—

Source: "Forest Ownership and the Case for Diversification," by Ken Drushka, in *Touch Wood: BC Forests at the Crossroads*, Ken Drushka et al (Madeira Park BC: Harbour Publishing, 1993), p. 4.

At the time Canada was settled and various colonial administrations set up, there was no resident population of established land owners as there was in Europe and other countries. Native inhabitants were either pushed aside or relieved of their land rights through treaties. Some areas were held by commercial interests such as the Hudson's Bay Company, which held Rupert's Land, and their claims reverted to Crown ownership when governing bodies were established. At about the time of Confederation, provincial governments ended up with ownership of and jurisdiction over all lands within their borders that were not held by the federal government or that had not already been granted to private owners. Native Indian reserves were held in trust by Ottawa. Apart from portions of Nova Scotia and New Brunswick, and much of southern Quebec and Ontario, that included most of the forest land in Canada. This was particularly true in BC.

There is no indication in the historical record that Canadians favoured government land ownership on principle. No one argued, for instance, that title to agricultural land should remain with the Crown. But in the late nineteenth and early twentieth centuries, the continental forest industry did not see itself as a permanent industrial fixture in local economies. Unlike the European industry it was transient, cutting down the forests and then moving on. If it was the local custom to acquire title to public land in order to obtain the timber growing on it, that was done. After it was logged, the land was sold to farmers for nominal sums or allowed to revert to the Crown in lieu of taxes.

As competition intensified and the purchase price of private forest land rose, it cost the forest industry more and more to obtain its timber this way. As markets in the US expanded after 1865, and Canadian lumber producers sought to compete in them, provincial governments offered leases and licences granting ownership of timber only, with royalties payable after it was harvested. Now lumbermen could avoid tying up scarce capital and paying interest charges and land taxes, which made it much easier to compete in US markets.

At the same time, Bernhard Fernow and other conservationists contended that private interests, unless they were large, permanent corporations, were incapable of managing forests. "Only governments and perpetual corporations or large capitalists can afford to make the sacrifices which are necessary to prepare now for such a management," Fernow told a Kingston, Ontario audience in 1903.[1] The facts did not support Fernow's statement, as he well knew. Most managed forests in Europe at that time were owned by individuals, family estates or small firms, as they are today. Fernow's own family owned a large forest estate in Prussia, and he had been educated in forestry so he could manage it. As a student, he adopted some of the current intellectual fashions and, abandoning his secure future, came to North America. He later wrote about "the ideal, the socialistic" state[2] and extolled the virtues of large trusts, which "may prove next to governments the most hopeful agencies for practising forestry, since they control large areas under uniform and continuous policy."[3] In Fernow's time there was no history of forest management by states or trusts. His opinions grew out of intellectual idealism, not experience.

Fernow's ideological views were widely accepted by turn-of-the-century conservationists and were central to the evolution of conservationist thinking in the US. Those views provided the rationale for the creation of US national forests, the first significant departure from the policy of private land ownership adopted at the time of the 1776 revolution. In Canada, with its colonial heritage, Fernow's ideas found even more fertile ground with both governments and the timber industry; the industry's enthusiasm may have been fuelled by the realization that government, not it, would be responsible for financing forest conservation.

With the establishment of provincial forest services in the early twentieth century, institutionalization of the state ownership policy began. From 1912, BC government agencies became the policy's most powerful advocates, and eventually its most ardent defender in BC would be the province's chief forester. The evolution of the Canadian corporate state, which acts to extend its own interests rather than those of its citizens, was not something imposed upon the institutions of government from outside. It was an idea that developed slowly from within, and one of the first principles adopted and then vigorously defended was state ownership of forest land and administration of it by agencies of the state.

The policy, which had been brought to BC from Ontario, was designed not to provide future opportunities for sound forest management, but to meet more immediate needs. Chief Justice Gordon Sloan acknowledged this in his 1957 report: "To conclude that the earlier settlers, who had a hand in framing the forest laws and policies of those days, were, in so doing, deliberately thinking in terms of generations ahead is, I think, with respect, to attribute to them an unwarranted prescience. Early administrations were, in my view, motivated by the immediate objectives important to the era in which they lived. They were confronted with the critical necessity of obtaining Crown revenue and of finding means to encourage the utilization of forest products."[4]

The question of who owned forest land was not much of an issue until the 1930s. By that time the province's first chief forester, H.R.

MacMillan, had built his lumber export firm into the largest in BC and was acquiring his own timber supply. His first major purchase was a block of private forest land in the Ash River valley, near Port Alberni. When logging commenced in this forest it was conducted in a manner to encourage natural regeneration of a new forest and to protect the new forest from fire and disease. It was one of the earliest examples of forest management in BC.

Meanwhile, in 1937, F.D. (Fred) Mulholland, a senior Forest Service staff member, went to Sweden and Finland to examine forest management practices and on his return wrote a report discussing the relevance to BC of what he had observed. Both countries were operating on a sustained yield basis, with managed young forests producing commercial timber at rates that he felt could be surpassed in BC. "On well-managed Scandinavian forests," he wrote, "a growing-stock of about 1,000 cubic feet per acre, producing an annual yield of 50 cubic feet, is considered fairly good. This is comparable with growth in the interior of British Columbia but on the coast a similar quality of wood can be grown at twice this rate because of the more favourable climate. An average site on Vancouver Island can easily be made to produce a yield equal to the best forests in Sweden."[5]

Mulholland was particularly interested in the social organization of the two countries as it pertained to forest ownership and the organization of forest workers.

> The majority of the forests are privately owned, chiefly by farmers. Wood is a cultivated product like agricultural crops . . . 76 percent of the forests in Sweden and 58 percent in Finland are privately owned. There are few large individual estates . . . The average farm in Sweden has 75 acres of forest. In Finland 96 percent of the farmers' forests are under 500 acres and 75 percent under 125 acres. Co-operation and decentralization of authority are the two most outstanding features of the Scandinavian democracies. There are basic laws which prohibit the devastation of forests under any ownership and forests whose owners have violated these

> laws are "closed." That implies that no further cutting is allowed until the area concerned has been rehabilitated by natural growth or planting. But the enforcement of these laws in private forests is entrusted to special organizations controlled by the owners themselves . . . I visited a number of forest-farms in Sweden and Finland and consider that their standards of comfort and culture are quite equal to those of our rural homes . . . Such people can withstand the vicissitudes of industry with less hardship than a landless labour class. It should not be impossible to develop a similar stable class of forest-farmers in BC . . . Vancouver Island should be especially capable of such development co-incident with the removal of difficulties caused by the heavy virgin stands and the type of industry necessary for their exploitation.[6]

Interest in forest management was on the rise, but it ended abruptly with the start of World War II. In 1941 BC's chief forester, E.C. Manning, was killed in a plane crash on his return from Ottawa while assisting in the war effort. C.D. Orchard, head of the Forest Service operations division and a career bureaucrat, was chosen to replace Manning.

Eighteen months later the new chief forester sent a confidential memo to the minister of lands, A. Wells Gray, whose portfolio included forests. It contained an eloquent argument for the adoption of sustained yield forest policies in BC, including a description of some of the administrative mechanisms required to implement these policies. One of these provisions was for a new form of tenure Orchard called a Forest Working Circle Reserve, which he later referred to as a Forest Management Licence (FML), and he included a draft of legislation providing for the licence in the Forest Act. It began: "The Minister, on the application and consent of any operator or intending operator, and in accordance with the further provisions of this part, may reserve any land, including timber licences, timber leases, pulp leases, pulp sales or timber sales, owned by the applicant, for the perpetual use and sole use of the applicant, hereafter called the licensee, and may add to such privately-controlled lands and similarly reserve sufficient additional Crown lands to establish a practical forest working circle for sustained-yield management."[7]

In other words, the holder of any form of tenure on Crown-owned forest land, or anyone else, could obtain permanent and perpetual control of that land, and enough additional land to provide an area of forest that would provide a timber supply to maintain this scale of operation forever. Title to the land would remain with the Crown, which would also oversee management of it. Through this form of tenure, Orchard believed existing forest companies could be induced to begin practising forest management which, in time, would establish a sustained yield for BC's forests.

The government's eventual response was to appoint Gordon Sloan to conduct a Royal Commission to, as Orchard later put it, "canvas the proposition, both for the value of his findings and as a measure of public education."[8]

Orchard served the commission as an advisor. During the course of the commission, the attention of the forest industry and the public was concentrated on the concept of sustained yield, and little attention was paid to his tenure proposal. "It didn't cause any great stir amongst [industry]," he commented later, "just another crazy Civil Service brain wave that wouldn't be heard of again."[9]

Sloan's report included a section on forest land ownership, and although he came to no conclusions, he did comment that "the present relation of Crown ownership to private ownership is the result of the long-continued and wise governmental policy of restricting the absolute alienation of Crown timber lands."[10]

He spent much more time on details about an aspect of the ownership question that may have had a greater impact on government policy—the relative revenues generated from privately owned and Crown forest lands.[11]

AVERAGE LOG PRODUCTION AND CROWN REVENUES—
1934–43

Tenure	% of annual cut	% of Crown revenues
Crown grants	40.3	4.1
Timber Licences	26.8	37.2
Timber Leases	16.4	16.9
Timber Sales	14.5	34.5
Timber Berths	20.5	4.7
Other minor tenures	1.5	2.6

Source: Gordon Sloan, *The Forest Resources of BC*, pp. 81–82.

Since the Crown received a disproportionally small amount of direct revenue from private land owners, there was little encouragement to allow this form of tenure to expand.

In his section on tenure, Sloan echoed Orchard's proposal for Forest Working Circle Reserves, suggesting "a form of tenure permitting the operator to retain possession in perpetuity of the land now held under temporary forms of alienation, upon condition that he maintain these lands continuously productive and regulate the cut therefrom on a sustained-yield basis."[12]

In fact, Sloan went a step further and advocated two kinds of working circles, public and private. The private ones would be those described by Orchard in his 1942 memo, designed to provide a cut sufficient to supply a company's mills. Public working circles would be managed either by the government to provide timber for market loggers, or by private owners without mills who would sell their timber on the open log market. Although working circles would remain in state ownership, they would be awarded in perpetuity.

Sloan's report appeared in December 1945, too late for amendments to the Forest Act to be included in the 1946 session of the legislature. Thus there was plenty of time to discuss which of his recommendations would be incorporated into law the following year.

On the question of sustained yield, there was virtually universal agreement. But an intense debate developed over tenure provisions. It was led by Mulholland and John Gilmour, MacMillan's chief forester. A serious breach had occurred between Orchard and senior members of the Forest Service, including Mulholland, and several of them resigned in 1946. The exodus took place partly because industry was hiring professional foresters, but there were other factors as well. Orchard was considered an autocrat, reflecting his ideas on how the industry's forest management activities should be supervised. There was also disagreement over tenure policy provisions in Orchard's draft legislation, including the question of land ownership.

In September 1946, Gilmour wrote to MacMillan that he and other foresters were getting nowhere with Orchard, and urged that forest company heads meet with Premier John Hart and Forest Minister E.T. Kenney to impress upon them that "no forest legislation, the work of one man [Orchard], should be placed before the Legislature before industry has had time to study it." He also pointed out to MacMillan the major weakness in the proposed new tenure as an instrument to encourage private companies to invest in forest management: "While a perpetual lease has almost all the advantages of a Crown Grant it lacks some of Crown Grant security."[13]

Mulholland, now employed as the Canadian Western Lumber Company's chief forester, endorsed these views in a confidential letter to John Liersch, who had left the Forest Service to teach forestry at UBC: "I do not think the Minister is hostile to private forestry; if he is a bit cynical about it and inclined towards State monopoly of forest ownership it is because his Deputy [Orchard] has not outgrown the Socialist theories which most foresters seem to pick up at College . . . The Government should, of course, continue to sell forest land for tree farming, as for other kinds of farming, to those who prefer this form of tenure to the proposed new leases."[14]

What Mulholland, Gilmour and others in industry were discovering was that regulations for the new Forest Management Licences would discourage their use by established forest companies, most of whom owned forest land. These companies were increasingly reluctant to place the management of their private land under control of the Forest Service in exchange for additional land they could lease under Orchard's new tenure. MacMillan noted: "As the proposed legislation is most important to the people of British Columbia, it would be a pity if it were set up on some theoretical basis as a result of which it would be a dead letter on the statute books so far as activities by companies are concerned. This would be a misfortune because the Government will have plenty to do in stocking and managing forest land and should enlist the help of private companies if possible."[15]

Even Orchard was not prepared to argue very strenuously for continued public ownership of all the forest land, although he did express the view in a 1946 meeting of Canadian foresters that "there seems to be no good reason why exclusive Government ownership and management of forest lands should not be successful and satisfactory." Then he went on to moderate his position: "It is my own opinion that Government is best able to own and manage forest lands for timber production, but that a reasonable measure of private ownership and management ranging up to, but not exceeding 50 percent, and preferably less than 50 percent, would introduce a healthy competition and act as a desirable tonic on management policy and progress."[16]

The following year Mulholland addressed the assembly on the same subject. After arguing persuasively that BC's policy of state ownership had not been maintained to conserve forests for perpetual production, he concluded:

> In past years there has been a good deal of propaganda designed to create the impression that without more Government ownership and control of forest lands, we are headed for a catastrophe. This is, I think, largely a reflection of similar propaganda in the United States. There it is understandable, with 80 percent of the forest land still in private ownership. In British Columbia with 90 percent in State ownership it would seem more reasonable to try to increase the interest of the private citizen in raising this crop so natural to the Province, and so important for its prosperity. At any rate no hindrance would be put in the way of anyone who wants to buy suitable forest land and who undertakes to keep it pro-

> ductive. Until the Government has its own tremendous forest estate in satisfactory condition it should welcome those who wish to relieve it of some of its burden, especially as private land will contribute to the Treasury through taxation while the trees are growing.[17]

Meanwhile there was more opposition to the Forest Management Licences proposed by Orchard and the government. Early in 1947 the BC Lumberman's Association, representing most of the big forest companies, presented a brief objecting to several provisions in the draft legislation. Independent market loggers on the coast, who had banded together in the Truck Loggers' Association, began to agitate against FMLs because they were convinced this tenure would eventually put them out of business.

The government did not relent. In the spring of 1947 it circulated copies of proposed changes to the Forest Act, allowing only two weeks for discussion. MacMillan wrote to the BC Loggers' Association: "It is not reasonable to expect the industry—who could not begin a study of the Bill until they saw it—to complete their thinking in two weeks . . . The only reason that could be brought forward for not giving everybody time to think would presumably arise from the desire that the public and the industry should not be given time to think."[18]

By the time the legislation passed, industry agreed the FML concept was seriously flawed and provided little incentive to practise sustained yield forest management. But their position was soon shaken by negotiations that were going on between the BC government and a group of US businessmen.

On January 22, 1947, Orchard and Kenney met with George Schneider, two other representatives of the American Celanese company and Bob Filberg, the president of Comox Logging, which ran one of the largest logging operations on private land near Courtenay. Although it would be another ten weeks before legislation enacting Forest Management Licences would be passed, Kenney had already promised the Celanese group a reserve on more than 2.6 million hectares of land on the Skeena and Nass rivers. The company, which had no established business in BC, intended to build a pulp mill in Kenney's riding, near Prince Rupert. Now they were back with Filberg asking for a reserve in another area, which included land already placed in the province's first public working circle along Johnstone Strait, north of Campbell River. The visitors intended to build a pulp mill at Duncan Bay, across the Strait of Georgia from the long-established Powell River Company pulp mill. This was too much even for Orchard, who described the incident later:

> This was a new proposal to join forces with Comox Logging, a proposal of which we had heard nothing to the time of this meeting. I refused to recommend to the Minister a reserve on 15 minutes' notice, much to the annoyance of Mr. Schneider who said Celanese didn't have to go asking for favours—they were too big—they would be more than welcome in Mexico—and more of the same tenor—none of which bothered either the Minister, or myself, very much. Schneider looked, and acted, like a spoiled brat, and after his tirade we still felt that we were entitled to a little time to consider a proposal of that magnitude, and declared accordingly. I spent all the next morning with [the Celanese forestry consultant] when we drew up a letter of application addressed to the Minister, which he took to Vancouver. I discussed the whole proposition with [senior Forest Service staff] during the afternoon and thereafter advised the Minister that it looked OK, but that an unannounced reserve to this new industry probably would result in violent protest from others, especially Powell River, with whom the new company would be in competition. The foregoing had no particular bearing on the course of the new legislation, but does indicate a keen interest in the matter once it appeared that the Government was seriously considering some such policy.[19]

Orchard could not have been more wrong in his concluding comment.

In the end Celanese was granted FML #1 at Prince Rupert, but did not get the reserve on Johnstone Strait. However, Filberg returned with another US company, Crown Zellerbach, which was granted a reserve on the same land that became FML #2. The consequences of these decisions were profound.

The forest industry was stunned when the reserves were announced. FML #1 had been granted to a foreign company that held no other tenures of any kind that would be supplemented with Crown land to form the FML, although Orchard had declared that to be the intent of the licence. The FML was for an enormous area, much larger than the modest-sized tenures Orchard had contemplated awarding to "several hundred" established logging firms. And, to add insult to injury, it was located in the minister's riding.

FML #2 covered more than 214,000 hectares. Its most offensive feature was that it included an area already set aside as a public working circle, which was to have provided timber for dozens of independent market loggers who sold their logs to several sawmills in Vancouver. They were allowed twenty-five years to harvest the remaining mature timber in the area. And while Comox Logging was a BC company with land to put into the licence, the money behind the new venture was from outside the province.

The message implicit in Orchard and Kenney's actions was that if established BC forest companies were going to oppose and boycott the new licences, the government was quite prepared to turn over the province's Crown-controlled forests to competitors from outside BC. As soon as the legislation was passed in April, established BC companies began filing applications for reserves, and another timber rush was launched.

The government of the day and its successor also made it clear that critics of their policies would never receive licences, as Sam Dumaresq, the second-generation co-owner of a prominent independent logging company discovered:

> We disagreed strongly with [the allocation of FML #2] and I spent most of one winter in Victoria working with Clair Smith

> of Smith and Osberg Logging trying to stop it. We were the first ones to publicly object to the Forest Management Licences, and in the end the Truck Loggers' Association supported us.
>
> We didn't get any farther than the front door. We met with the Cabinet and presented a brief written by Clair, but they did not pay much attention to us. Later on, I came down on the train from Kelowna with W.A.C. Bennett and tried to explain our objections to him. We had started spending summers in Kelowna and I got to know him there. He was the premier then, but he wasn't listening to any objections about this.
>
> During this time the control of the industry moved from family businesses into big public companies. I think the politicians were so completely dominated by the big companies. For self protection we eventually applied for a Forest Management Licence, but we stood the same chance as a snowball in hell because our name had appeared on this protest to Cabinet. I don't think Clair Smith would have got one either. A few years later we decided to get out of the lumber business.[20]

By their actions, the minister and the chief forester had undermined their own legislation. Instead of acquiring licences to practise forest management, companies snapped them up as a means of securing mature timber. Orchard himself acknowledged this in his unpublished memoirs.

> Whereas I had thought that, given the authority, we just might induce some public spirited and far sighted operator to take up a forest management licence with all its attendant responsibilities, the fact turned out to be that almost at once we were deluged with applications. Industry saw in an assured timber supply a chance for capital gain that I had quite overlooked, and which no one in Government or Civil

> Service detected. Regionally, and particularly on the south coast, timber already was in short supply. Companies that operated on a hand to mouth basis clamoured for an assured timber reserve safe from competitive sale.[21]

These observations were typical of Orchard's attitudes toward private sector forest companies. He considered them all to be greedy, voracious pillagers of the forest. Of course they were aware of the potential for windfall capital gains in the new licences, but what Orchard left out of his personal account is a description of the pressure that Kenney, and possibly Orchard himself, applied to the companies. In effect the established operators were told to accept and apply for the new Forest Management Licences or they would not be able to obtain Crown timber through other types of leases and licences. As they were hearing this, FMLs were being granted to competitors within their operating areas.[22] To protect themselves, they fell into line and submitted FML applications.

While the FML system was flawed from its inception, it took the election of the Social Credit government in 1952 to set the problems in stone.[23] To Orchard's bureaucratic ineptness and Kenney's pork-barrel style of politics, the Socreds added outright corruption. Their forest minister, Robert Sommers, simply sold FMLs under the table for pitifully small bribes, and eventually ended up in jail for his transgressions.

Even these events, however, could not dislodge FMLs from the list of tenures available. Before Sommers was convicted, Sloan had conducted his second Royal Commission, in 1955, which turned into a kind of public trial of the FML system. By then, fourteen of the licences had been issued, covering 3.8 million hectares of forest land, and applications for dozens more had been filed with the Forest Service. Those who had already obtained licences fought to keep them; those with applications pending fought to get them.

The chief opponents of the system were small independent market loggers on the coast who bought timber via Timber Sales, at competitive auctions. They realized the FML system would eventually tie up the best areas of Crown-owned timber, squeezing them out of business and leaving forests under control of a few large corporations. Curiously, the most eloquent articulation of their argument came from the owner of the biggest forest company in BC, H.R. MacMillan. For several days the seventy-year-old former chief forester argued in front of Sloan against the allocation of any more FMLs in the most important area of the coastal region, even offering to withdraw his company's application. "It will be a sorry day for British Columbia when the forest industry here consists chiefly of a very few big companies, holding most of the good timber—or pretty near all of it—and good growing sites to the disadvantage and early extermination of the most hard working, virile, versatile, and ingenious element of our population, the independent market logger and the small mill man."[24]

MacMillan was attacked vigorously by representatives of the other big companies, who were allowed the right of cross-examination. He fought back in ringing phrases that still echo through the industry, and ended with a dramatic prediction of what the FML system would lead to: "A few companies would acquire control of resources and form a monopoly. It will be managed by professional bureaucrats, fixers with a penthouse viewpoint who, never having had rain in their lunch buckets, would abuse the forest. Public interest would be victimized because the citizen business needed to provide the efficiency of competition would be denied logs and thereby be prevented from penetration of the market."[25]

MacMillan's prediction turned out to be devastatingly accurate, but not just because of FMLs. In addition to Orchard's private working circles in the form of FMLs, public working circles called Public Sustained Yield Units (PSYUs) had been set up. These were to be managed by the Forest Service, to provide timber for market loggers via competitive Timber Sales. At the time, small loggers were assured repeatedly that these provisions would protect them from the larger companies, almost all of which had obtained FMLs. But nothing in the Act prevented FML licensees from operating in the PSYUs, and within a short time the big companies were also bidding on Timber Sales there.

In addition, the new act's sustained yield policies called for the adoption of cut controls in the working circles, setting maximums and minimums to be harvested each year. In those PSYUs that were already being logged at levels close to or exceeding the cut limits, intense competition for timber developed and the established operators in the PSYUs found themselves being outbid by newcomers. To protect their position, the Forest Service adopted an informal quota system that gave established operators annual harvesting rights proportionate to their cutting rates in each PSYU prior to the establishment of its cut limit.

This quota system had no basis in law; Peter Pearse said later that it had actually violated the law.[26] Over time it evolved into a powerful tool in the hands of the forest minister to give big companies control over most cutting rights in PSYUs. Quota became saleable, and through one means or another small loggers relinquished their rights. On the coast, some were simply outbid by the big operators during periods of low log prices or squeezed out during rigged bidding procedures. Others were induced to exchange their quota positions for what looked at the time like lucrative logging contracts or cash payments. In the Interior, where most quota was held by small sawmill operators who did their own logging, the consolidation of cutting rights in the hands of fewer companies was not so contentious. In the end, however, the result was the same—the consolidation of control over the timber supply by relatively few companies. What made this consolidation possible was the fact that the land and timber were state owned, and very little money had to be paid to acquire cutting rights. Combined with complicit bureaucrats and governments, the process was much less difficult and less expensive than it would have been in a country such as the US or Sweden, where most of the forests are privately owned.

In 1958 the Social Credit government renamed FMLs Tree Farm Licences (TFLs) and acknowledged that their real function was to allocate a secure timber supply. The government repealed the perpetual terms of these licences and established a new term of twenty-one years. By then, twenty-three TFLs had been granted. The companies holding them naturally took the position the legislation applied only to subsequent licences, while the Forest Service tried to apply it retroactively. Either way, the amendment dispelled any notion that TFLs provided a suitable tenure for long-term private investment in forest management. This idea had already been abandoned in practice: when licensees spent money on silviculture, they were either reimbursed for it, granted increased allowable annual cuts, or both. By 1960 the government was, in one way or another, footing the bill for forest management on the TFLs—which were not a forest management tenure, but a harvesting tenure.

By the late 1960s, with a large portion of quota in most PSYUs in the hands of relatively few companies, the government introduced another form of tenure acknowledging this concentration and strengthening it further. Timber Sale Harvesting Licences offered those holding several Timber Sales within a PSYU the opportunity of consolidating their rights in one licence, and extending its term from the usual three years to ten. These licences were scooped up quickly. By 1974 they accounted for 60 percent of timber harvested in the PSYUs.

Another tenure innovation was introduced by the Forest Service in the 1960s, this one arising from application of the yield regulation system. Cutting levels within the PSYUs were established on the basis of how many trees found in the PSYU were used. What was left lying on the ground, as long as it was not considered merchantable timber by these standards, was not counted. By the 1960s, logging and milling technology had evolved to the point where it was economical to use much of this waste timber. Companies that could demonstrate an ability to use it were therefore granted increases in their quotas. These volumes, and additional excess volumes within the PSYUs, were awarded in the form of Timber Sale Licences known as third band sales. As was the case with many such "minor" adjustments to the rules, this one soon took on a life of its own. When Peter Pearse examined the industry a decade later, third band Timber Sale Licences accounted for 36 percent of all timber cut in the PSYUs, with the highest concentration of them found in Interior forests.

By the time Pearse held his hearings, the controversy over TFLs had subsided. Many of them were being managed far better than PSYUs under the care of the Forest Service, which was hampered by low budget allocations. Now, in 1976, the major concern was with how PSYU cutting rights had become concentrated in the hands of relatively few companies. One measure of their control was that the Timber Sale Harvesting Licences and third band Timber Sale Licences, used almost exclusively by the big companies, took up 85 percent of the allowable annual cut in the province's PSYUs.

Changes in tenure adopted in 1978, after Pearse submitted his report, served in many ways to consolidate the status quo. Quota positions of established companies were formalized with the creation of Forest Licences, which superseded existing Timber Sale Licences, including the third band variety, and Timber Sale Harvesting Licences. Forest Licences provide a specified volume of timber each year, for a fifteen-year term, with an "evergreen" renewal clause providing five-year extensions. Old temporary tenures, many dating back to the last century, were converted into new Timber Licences, which could be rolled into existing TFLs after they were harvested.

Timber Sale Licences, permitting sale of timber at auction, were retained as a tenure device to allocate timber under a new Small Business Enterprise Program (SBEP), the Social Credit government's answer to Pearse's strongly worded warnings and complaints from small loggers. Bidding on these sales was restricted to certified small firms. The forest minister of that period, Tom Waterland, extravagantly promised to allocate 25 percent of the annual harvest to this program. The SBEP also included the new Woodlot Licences, replacing and broadening the scope of a little-used tenure, Farm Woodlot Licences.

Four years later the government quietly passed an amendment allowing holders of old Timber Sale Harvesting Licences and Timber Sale Licences, and new Forest Licences, to convert these volume-based tenures into area-based Tree Farm Licences. Little use was made of this provision until Forest Minister Dave Parker's disastrous attempt to apply it on a wholesale basis in 1989.

Essentially the same tenure system is in effect today on state-owned forest land. In addition there is about 2.5 million hectares of private forest land in the province, 919,005 hectares of which is classified as Managed Forest Land (MFL) and is taxed at a lower rate in exchange for the owners' commitment to manage it sustainably. Most MFL is owned by the big companies and most of their holdings are included in Tree Farm Licences. About 20 percent of it is held in small parcels of less than 1,000 hectares and is managed under the direction of the BC Assessment Authority.

The other 1.6 million hectares of private land is owned by more than 20,000 individual owners and is known as unmanaged forest land. It may consist of part of a ranch, the back corner of a farm, recreational property, or simply forested land not used for any particular purpose. Much of it is badly mistreated, neglected or in the process of being deforested because there are no incentives for its management or regulations affecting the fate of the forests growing on it.

VOLUME OF TIMBER HARVESTED BY TENURE 1996–97

(Thousands of cubic metres)

Crown Land		
	Tree Farm Licence	12,481
	Forest Licence	39,240
	Small Business Enterprise Program	7,967
	Timber Licence (in TFL)	2,195
	Timber Licence (non-TFL)	899
	Woodlot Licence	361
	Other (Beachcomb, misc)	3,562
	Total from Crown land	66,707
Private Land		
	Within TFL	328
	Outside TFL	6,786
	Total from private land	7,114
Federal land and Indian Reserves		191
	Total for BC	74,011

Source: Ministry of Forests Annual Report, 1996–97, p. 97.

Under the tenure system that has prevailed for the past twenty years, the concentration of control over cutting rights that Pearse warned British Columbians about has continued unimpeded. Instead of maintaining and strengthening a diverse industry, tenure policies have fostered regional timber monopolies, which have in turn concentrated control over the various processing sectors. The tenure system has been the primary factor in the development of a monolithic forest industry, as susceptible to destruction by natural economic events as is a monocultural forest to an insect or disease attack.

The worst failing of the tenure system, however, is that it has done nothing to encourage a healthy forest management capability in BC. In fact, it has impeded that capability, and until it is revised all other attempts to solve the problem will likely fail.

In 1994, Peter Pearse and Daowei Zhang examined silvicultural expenditures on more than 2,000 forest areas in the BC southern Interior and on the coast. These areas were covered by four different forms of tenure: private land in TFLs, Crown land in TFLs, Timber Licences and Forest Licences. Forest Licences are for fifteen-year terms, Timber Licences for variable terms, Crown land in TFLs for twenty-five years, and private land in perpetuity. Pearse and Zhang found a direct correlation between the amount spent on silviculture and the security provided by the different tenures. Between 1987 and 1994, for every $1.00 spent on Forest Licences, $1.14 was spent on Timber Licences, $1.24 on Crown lands within TFLs, and $1.73 on private lands within TFLs. They concluded: "The policy implications of these findings are significant. Forest Licences—the least secure form of tenure, associated with the lowest silvicultural investment—is by far the most important, accounting for well over half the timber harvested in the province and applicable to an even larger proportion of the province's forest resources. If the silvicultural investment observed on private lands can be assumed to be efficient and beneficial to the owners, the much smaller investments under Forest Licences means that substantial opportunities for beneficial silviculture are being lost on public forest lands."[27]

Another disadvantage of the existing tenure system is that it requires a high level of administration by the Forest Service. The system was developed just after World War II, a time when state bureaucracies in Canada were expanding rapidly. Now we may have reached or even exceeded the limits of what we are willing to support in the way of state activity: governments that have reduced spending and downsized their bureaucracies are more and more in favour. Taxpayers are apparently not willing to maintain the existing state superstructure, let alone expand it to the levels required to properly manage the country's state-owned forests.

Under certain circumstances there are advantages to state ownership of forests. A large portion of BC forests are low in commercial value: much of this land is now in parks, wilderness reserves, watersheds and other forms of protected status, and no benefit would be gained by transferring ownership of these areas into the private sector.

Other arguments for state ownership are harder to support. For example, state ownership is said to allow government to direct and influence economic development within the province. This may have been so in the past, when there were large areas of unpopulated, undeveloped land in BC. But today, with BC's many diverse regional and local economies, a more decentralized and flexible ownership pattern and administrative structure are more appropriate. Another argument is that if forest land were privately owned, most of it would eventually end up in the hands of a few large forest companies. There is no evidence to support this argument. In every country in the world with substantial areas of private forest land, most of it is owned in small holdings and has stayed that way for many generations. Agricultural land in Canada remains in the hands of a multitude of small private owners and there is no trend toward a corporate takeover of farms. In fact, what has occurred in Canada and especially in BC, with our habit of leasing state-owned forest lands to the private sector, is that a few large corporations have obtained control of these leases. In virtually every Canadian jurisdiction with large areas of state-owned forests, governments have been complicit in allowing large corporations to assume control of their use.

The BC forest industry today is characterized by big corporations, big unions and a big state bureaucracy. The industry did not become what it is because the citizens of the province wanted it this way. It evolved largely as a result of policies designed by bureaucrats and implemented by politicians who were not always working on behalf of their constituents. As submissions to the Pearse and Forest Resources commissions indicated, the people of BC have no great desire to maintain this type of social and economic structure. What citizens have called for, repeatedly, is a more diverse structure in which a much larger number of smaller, more flexible enterprises can function alongside the big companies.

Until we reorganize the system to be more diverse, we will also have to continue grappling with the tension between BC's urban and rural populations. Almost all the large corporations, unions, government departments and environmental organizations are located in the cities, while those firms located in outlying centres and rural areas tend to be smaller. The past fifty years have seen a steady migration from rural to urban centres, coincidental with the concentration of resource rights. This migration has not occurred because people developed an overwhelming urge to live in cities, but largely because they are increasingly unable to build local economies on a resource base they can count on. Government action and inaction have contributed to the migration because it is more convenient to provide services to cities than to rural areas. Corporate bureaucrats, too, prefer not to rely on an indigenous labour force. As one coastal corporate forester told me in the early 1970s: "We don't want a local workforce. Someone will always want to take the day off to fix his chicken house. We want workers living in camps, where we can control them for ten days at a time."

During its 1991–95 term in office, the NDP government avoided the question of forest tenure. When asked about the subject, Forest Minister Andrew Petter said tenure would be dealt with during his government's second term. When Glen Clark took over as premier, he said no changes would be made to the tenure system under his administration. As a graduate student at UBC's school of community and regional planning in 1985, Clark wrote a master's thesis on BC's forest tenure system indicating he understands clearly the central role it plays in the provincial economy and the power it provides for state intervention in the economic and social life of the province. A decade later, Premier Clark made it clear he was not willing to abandon a system providing the government with such a powerful tool to regulate the lives of its citizens, even though that system fails to provide either a healthy forest economy or sound forest ecology.

In 1999, however, an exception to the premier's rule was made. The government proposed to compensate MacMillan Bloedel and other companies for timber lost to parks with title to Crown land. The reason offered for this inconsistent application of tenure policy was the need to save government money, but the sums involved were not large enough to warrant such a departure from established policy. The real reasons stemmed from the hostile investment climate the government's policies had fostered, and the likelihood that, if compensated in cash, some companies would choose to reinvest outside BC.

If it is accepted that a healthy forest economy based on ecologically sound principles of forest stewardship is a desirable goal for BC, then the forest tenure system must change. Nowhere in the world has any forest tenure system based on state ownership of forests achieved this objective. And BC's experience over the past fifty years demonstrates that the system in place does not enable us to achieve this objective.

Other forest nations have shown that private ownership of forest land can lead to good forest management and the creation of vibrant forest economies, without a detrimental effect on other forest values. Private ownership of forest land is not a goal or an objective in itself; it is merely a means to realize the social objectives of the people of BC. The existing tenure system is an obstacle to reaching these goals. It must be changed.

7 Revenues and Tariffs

Because most timber sold in British Columbia is from state-owned forests, and only a relatively small amount from private forests is sold on the open market, timber revenues are assessed and collected largely by government policy. The focus of these policies has changed over time, although this is not always obvious because official policies are not always the same as actual ones.

Timber revenue was first collected in BC in 1858 for the sale of forest land and the timber on it. The price was only 10 shillings an acre. In 1865 the colonial government of Vancouver Island issued a land ordinance that established the right to harvest timber from Crown land without acquiring title to the land. The ordinance provided for the collection of rents and royalties "as shall seem expedient to the Government." These well-established colonial fiscal instruments had been used in eastern Canada long before they were applied in BC.

Until 1888, rentals ranged from 5 to 10 cents an acre and the government charged royalties of 50 cents a thousand board feet. After that year the rates increased slowly, and by 1974 rentals on leases issued in this period and still in good standing had reached 50 cents an acre. After 1897, to encourage milling in the province, non-mill owners paid an additional 5 cents rental per acre. The element of competition was added in 1891 when timber leases surveyed by the government were put out to tender, and the following year competitive bidding on leases became mandatory. Pulp leases were issued between 1901 and 1903 to encourage establishment of a pulp and paper industry, and carried reduced rents of 2 cents an acre and royalties of 25 cents a cord. The various types of leases issued in this era were intended to provide timber to mill owners, and charges levied on timber reflected the government's policy of encouraging manufacturing in the province.[1]

To make timber available to independent loggers, the government designed a series of licences beginning in 1884. Pricing policy on these licences was set for the purpose of raising revenues for the government, and the rental and royalty rates were higher than on leases. The first licences were sold for $2.50 an acre, annual rentals were $10 and royalties were 15 cents a tree plus 20 cents per thousand board feet.

In 1901 the government obtained only 7 percent of its $1.6 million annual budget from timber revenues. Four years later, the expansionist government of Richard McBride modified the terms of Timber Licences so they could be sold, and royalties were adjusted annually to give the government a share of any increased value in the timber. Fifteen thousand of these new Special Timber Licences were issued in the following two years, before they were cut off. By 1908 timber revenues accounted for 40 percent of the province's $6 million budget. Expenditures on forests that year totalled $49,747.[2] A new forest revenue policy that essentially stripped the forests to finance government expansion became firmly established in a few brief years. It is a policy that has been used by provincial governments ever since.

After 1907, no more leases or licences were issued, and until 1912 there were no further allocations of Crown timber in BC. The first Forest Act, passed that year, allowed for a new timber tenure, Timber Sales, which reflected a new pricing policy. Initially Timber Sales were subject to rent and royalties, but these charges were soon abolished. The innovative levy was a "stumpage" charge. Originally this term referred to the value of a tree before it was cut, and took into account the cost of falling the tree and getting it to market. When used by private timber owners, stumpage refers to the value of their forest, or the price someone will pay them for it. But now, as used by the newly formed BC Forest Branch, the term took on a different meaning.

The government established an appraisal system for estimating the value of timber and subtracted from it the estimated cost of logging. This provided an "upset price," the lowest price at which bidding could start. The stumpage fee paid consisted of the upset price plus the winning bonus bid during auction.

The use of Timber Sales increased steadily, while the annual harvest from pre-1907 leases and licences declined. By 1934 Timber Sale stumpage fees were providing 14 percent of $2.3 million in forest revenue. By 1955 stumpage fees made up 88 percent of the $24 million forest revenue account.[3] That year $9.6 million was spent on BC forests.[4] After World War II, with the creation of Tree Farm Licences and widespread replacement of Timber Sales with Timber Sale Harvesting Licences, the proportion of timber acquired by competitive auction declined dramatically. Timber was sold through these new tenures at appraised stumpage rates. Through manipulation of the figures used to calculate harvesting costs and timber values, appraised stumpage rates became progressively undervalued, and by 1982 stumpage had fallen to 55 percent of forest revenue. That year the Forest Service cost $189 million more to operate than was collected in revenue. The net loss on administering the province's forests continued until 1988, by which time the accumulated "loss" on operating the Forest Service was $1.1 billion.[5]

An examination of stumpage rates paid by loggers obtaining timber through Small Business Enterprise Program Timber Sales shows the extent to which appraised stumpage rates undervalued timber. In 1982, about 10 percent of the Crown timber harvest was sold by this method, which requires competitive bidding. That year the appraised value of the 2.5 million cubic metres cut by SBEP loggers was $3.08 million. At auction this timber sold for $9.38 million—an average of $3.75 a cubic metre. The average stumpage rate paid for the 45.3 million cubic metres of Crown timber cut that year by the major forest companies, all of which was sold at the appraised rate, was $1.79 a metre.[6]

Part of the reason for the discrepancy was that SBEP loggers had fewer responsibilities than the companies obtaining their timber at appraised prices. But more important, over the previous few years, one step at a time, the appraisal process had been corrupted. Originally the process was intended to arrive at a rate close to the rate that would exist in an active timber market. To calculate this rate, accurate estimates of logging costs and log prices had to be made. On one hand, the system for reporting prices received for timber had been rigged to indicate unrealistically low values. On the coast, most log transactions were trades rather than sales and purchases, so it was a relatively simple matter to assign a low dollar value to the traded logs. On the other hand, procedures used to calculate harvesting costs had been manipulated to produce extravagant results. The Forest Service, with no direct experience in logging or log sales, ran the system, and over time its rates ceased even to approximate real values.

Furthermore, the system as a whole did not encourage efficient logging because if the logging industry did become more productive, it would only have to pay higher stumpage rates. Essentially the process estimated stumpage as the value obtained by subtracting logging costs (and a percentage of those costs as a profit and risk allowance) from the selling price of logs. In reality, this was a cost-plus system of logging: the higher the cost of logging, the bigger the profit allowance to loggers. An individual logger could gain by trying to be more efficient, but the industry as a whole had a built-in incentive to raise costs, or at least to refrain from resisting certain cost increases.

An example of this incentive, and of how the appraisal system distorted the economic basis of the industry, was its effect on wage rates. For many years, until the late 1980s, whenever the IWA pushed hard for wage increases, the big, unionized companies readily gave in rather than risk a strike. The increase in labour costs would then be factored into the appraisal process and be paid for in the form of lower stumpage rates. In effect, the government was paying for these wage increases. BC loggers became the highest paid in North America, if not the world, with no incentive to be any more productive. This, in turn, increased the need for mechanization of logging. To be able to pay these higher wages it was necessary to provide workers with machines to increase their productivity, which in turn led to a reduction in the logging workforce.

By 1982 the economics of logging in BC had become so distorted there were few people left in the province, apart from a handful of market loggers, with any idea about the real costs and benefits of the business. Then two events woke everyone up. The first was a global recession, which weakened the export markets into which most of the timber cut and processed was ultimately sold. The second was that competitors in the biggest BC market, the US, blew the whistle on the BC stumpage system. The Americans launched the first of a series of legal actions against various aspects of BC's forest administration system. Fifteen years later these actions had raised stumpage rates enormously and locked the industry into a new regulatory bureaucracy.

In 1987 BC responded to US actions by adopting a new appraisal system known as a "comparative value pricing" system. In this system the government establishes a "target revenue," or the total amount it wants to collect in the coming year from stumpage and royalty payments. Lumber price indexes and other statistical data are used to arrive at this target, but the major factor is the need to convince the Americans BC is not subsidizing the forest industry.

To calculate stumpage rates, the first step is to factor out royalty payments and treat them separately. The next step is to divide the remaining target revenue by the expected volume of timber to be logged, which yields an average stumpage rate. Then a stumpage rate is determined for each stand of timber logged. This is lower than average if that timber is less valuable, higher if it is more valuable. This is known as the "water-bed effect." If the rates go down for one stand, they will go up for another and the target revenue must be collected. Unfortunately, the water-bed effect produces unexpected ripples: if someone does not log their allocated timber volume, stumpage rates will rise on the timber that is logged. In 1997, for instance, many companies were unable to log their timber quota because they could not obtain Forest Service approval in time, or they had suspended logging during a dispute with environmentalists, or they had to postpone logging a stand because costs and stumpage fees were too high. Stumpage prices rose for the timber they did log, making it uneconomical for even more stands to be logged.

Stumpage rates are revised every three months, as target rates are adjusted in response to changing lumber prices and market conditions. In 1994, when various factors, including a rapid increase in prices, led to an increase in BC lumber shipments to the US, it was clear that a major stumpage increase was required to forestall another American legal action. After meeting for many long days under the umbrella of the Forest Sector Strategy Committee, government, industry and the forest unions agreed to a massive increase of $600 million a year in the target revenue. This amount became known as Super Stumpage, and most of it was allocated to Forest Renewal BC, purportedly for forest enhancement.

By the time the new stumpage system was fully operative, stumpage fees had soared, at least in comparison to those charged between 1982 and 1988.

STUMPAGE, ROYALTIES & RENT: 1986–1998

Year	Revenue	Crown timber cut	Revenue/m^3
1986	$173,000,000	73,000,000 m^3	$2.37
1987	$386,000,000	80,000,000 m^3	$4.83
1988	$627,000,000	77,400,000 m^3	$8.10
1989	$648,000,000	78,000,000 m^3	$8.31
1990	$605,000,000	70,300,000 m^3	$8.61
1991	$574,000,000	68,600,000 m^3	$8.37
1992	$616,000,000	67,700,000 m^3	$9.10
1993	$849,000,000	71,500,000 m^3	$11.87
1994	$1,430,000,000	66,400,000 m^3	$21.57
1995	$1,798,000,000	67,000,000 m^3	$26.84
1996	$1,733,000,000	67,600,000 m^3	$25.64
1997	$1,773,000,000	60,800,000 m^3	$27.36
1998	$1,415,000,000	56,800,000 m^3	$24.91

Source: Price Waterhouse, *The Forest Industry in British Columbia*, various editions.

At the same time Super Stumpage was levied, the government changed the way it determined royalty rates. Royalties on outstanding leases and licences had been edging up since the turn of the century, as timber values increased, but had always lagged behind stumpage rates. The rationale was that holders of these tenures had paid the carrying costs of owning them and therefore should not have to pay as high a rate as for timber cut from tenures that did not incur these carrying costs.

Other factors affected the decision to raise royalty rates in 1994. One had to do with the water-bed effect of the new stumpage system. Because royalties figure in the target revenue, raising royalties would push down stumpage rates. By 1994, royalty-bearing timber was owned by only six companies—about 60 percent of it by MacMillan Bloedel, whose competitors were enthusiastic about obtaining reduced stumpage costs at MB's expense. The government concurred, and decreed that royalties would be increased to 60 percent of stumpage immediately, and to 100 percent by 2001. MB and one other company, Timberwest, sued the government for $240 million over the issue; in late 1999 the matter was still winding through the courts. Another reason the provincial government was eager to raise royalties is because royalty payments are deductible from income tax. Raising them transfers revenue from the federal to the provincial government.

While stumpage and royalties are the major sources of government revenue from the forest industry, they are by no means the only ones. In recent years these payments have comprised about two-thirds of all payments to government, with various taxes making up the balance. Another large contribution to state revenues comes from income taxes and other payroll deductions from forest industry employees, an amount approximately equal to stumpage and royalty payments[7] (see Appendix, Tables 7A and 7B).

The changes in stumpage policy introduced since 1991, when the NDP came into power, are a reversion to those adopted by the McBride government in 1905—using the forests to finance an expansion of government activity. Ostensibly, the purpose of the Super Stumpage increase was to finance an enhanced silvicultural program through Forest Renewal BC. But in fact, FRBC money simply replaced other silvicultural funding sources, such as the federal-provincial Forest Resource Development Agreement. Instead of spending more on silviculture, the government off-loaded $100 million in annual ministry silvicultural costs onto FRBC, which then directed a portion of its budget into silvicultural projects employing displaced forest workers. The process transferred a substantial portion of silvicultural work performed by seasonal silvicultural contractors to displaced union workers. Silvicultural efficiency and productivity were reduced, and there was a net reduction in silvicultural accomplishments.

NDP stumpage policy maximized forest revenue to finance an expanded program of unemployment relief, retraining and relocation for unionized forest workers. This is quite different from the Social Credit policy of the 1980s, which was to minimize stumpage and maximize profits for the large integrated forest companies. In the end, both governments used the forests to keep their friends happy. Ironically, the one factor that kept both governments even marginally honest was the US forest industry.

The softwood lumber dispute between Canada and the US began even before the BC forest industry. Ontario sawmills began shipping lumber into the US market in the late 1850s, aided by generous tenure and revenue policies. American mill owners soon persuaded the US government to impose a tariff on Canadian lumber entering the US. In the 1880s, as the Ontario forest industry migrated west along the north shore of the Great Lakes, loggers began towing booms of logs out of Georgian Bay across Lake Huron to mills in Michigan. In response, the Canadian government imposed a series of export duties on logs to encourage the development of mills in Canada. In 1890, after intense lobbying in Ottawa and Washington by Ontario mill owners, the log export duty was eliminated in exchange for ending the US tariff on Canadian lumber. Then, in 1897, the US imposed the so-called Dingley tariff, a stiff tariff on lumber of $2 per thousand board feet, with provisions to raise it by an amount equal to any Canadian export tax on logs. In response, the Ontario government prohibited the export of logs.[8] BC followed this lead in 1901 with a law requiring that all timber cut from Crown lands be milled in the province. Five years later another law extended this requirement to timber cut on Crown land granted after 1906.

In the early years of this ongoing dispute, most of the moves were made by the US, Canadian and Ontario governments. BC producers were not a significant factor in the North American lumber trade. But as trade between BC mills and US buyers increased, BC operations came under closer scrutiny by Americans. Within the US there was then, as now, much disagreement over lumber tariffs. Home buyers, farmers, builders, building supply retailers and others were generally opposed to a tariff; lumber producers favoured it. Tariffs ebbed and flowed depending on the political fortunes of these interest groups. The end of World War I, for example, saw strong opposition to a tariff among survivors of the conservation movement, including Gifford Pinchot, who saw the import of lumber as a means of prolonging the life of US forests. Supporters of tariffs pointed to two factors that gave Canadian lumbermen unfair advantages in the US market: lower production costs due to lower stumpage and royalty rates, and the low value of the Canadian dollar.

The post-war shift in control of BC lumber export sales from US brokers to Vancouver exporters, such as H.R. MacMillan, undercut the influence of another US group strongly opposed to lumber tariffs. Exporters like MacMillan developed their businesses carefully, with an eye to maintaining the support of key US interest groups. When BC lumber mills created a co-operative marketing agency, Seaboard Lumber Sales, in 1928, this US support was eroded; perhaps for the first time, BC producers were instrumental in the decision to impose a tariff. A few years later MacMillan attempted to warn Seaboard about the risks involved in its move into British markets:

> I am somewhat governed by the fact that when the Seaboard Sales was put together some years ago to handle the Atlantic coast market, it seemed to be just as good as the idea now being adopted for the sale of lumber in the United Kingdom. Nevertheless, as the administration of the Seaboard Sales pro-

> ceeded, the management, as the American market began to weaken, continuously forced all possible Canadian lumber into that market, thereby incurring the growing enmity of the American lumber producers.
>
> Although when the business was first developed by the Seaboard Sales in the Atlantic Coast market, the dealings were almost entirely with wholesalers, the policy was gradually changed to a point where wholesalers were eliminated to a great degree and sales were made direct to most of the wholesalers' good customers. This finally earned for the Seaboard Sales the hostility of the wholesalers.
>
> As the strength of the Seaboard Sales developed, the management discovered in their hands a great power in control over freight, which power they used almost entirely against the interests of the Conference American Intercoastal Lines by continuously chartering below Conference rates and making almost negligible employment of American vessels, thereby bringing the American Shipowners' Association into the field as strong and aggressive enemies.
>
> The combination of these circumstances, so far as I have been able to learn during the past few years from my acquaintances in the shipping industry and the lumber industry in the United States, resulted in the imposition of the present prohibitive tariff on lumber into the United States.[9]

World War II put an end to lumber tariff fights, and during the expansionist 1950s tariffs did not become an issue. But by 1962 the US economy was in a slump and its lumbermen were calling for a tariff. Pat Carney, then working as a journalist, wrote in *Truck Logger* magazine:

> This fight is not based on sweet reason, but on politics, pure and simple. Back scratching, log rolling practical politics. And politicians are not much interested in such high-toned matters as balance of payments problems and trade liberalization. They are interested in votes.

> Consider the situation in which [US] senators like Warren Magnuson and Wayne Morse find themselves, as we hear it. They face important elections this fall. There is unemployment in their local lumber industries. Mill towns show the dismal signs of economic slowdown; closed plants, layoffs, "For Rent" signs in office windows.
>
> Meanwhile, ships stacked high with lumber sail out of BC ports. BC's share of the biggest US market, the Atlantic seaboard, climbs from 23 percent to 72 percent in five years. BC is a handy villain.[10]

Carney was right. The tariff issue has always been one of politics, pitting Canadian forest policies against US forest policies. In this case the BC industry recognized it and undertook to muster support for its position among Americans opposed to the tariff. Under the astute direction of Bert Hoffmeister, a Canadian war hero who later served as CEO of MacMillan Bloedel, the campaign was a success and no tariff was imposed.[11] Soon after, the markets picked up and the US lumber industry lost interest in a softwood lumber tariff until the next recession.

Twenty years later, with lumber markets sagging and BC lumber shipments across the border soaring, the Americans set off another tariff fight, first with a duty on shakes and shingles, and then with a request to the US International Trade Commission for a countervailing duty on Canadian lumber. Under the direction of the Council of Forest Industries (COFI), Canadian—primarily BC—producers insisted that forest policies in this country do not constitute a subsidy to industry. To strengthen their case, COFI, the Ministry of Forests and the IWA provided US investigators with information, much of which, to say the least, skirted the truth.

The Americans were unsuccessful because of a technicality, which was removed by new legislation in 1985. The following year, working through the well-funded Coalition for Fair Lumber Imports, they returned to the Commission with another request, which the US government agreed to consider. They pointed out that in spite of record high lumber demand, 630 softwood sawmills had closed in the US since 1977, while 85 new ones had opened in Canada, and 30,000 US mill employees had been put out of work. Since 1975, the Coalition argued, Canada's production of softwood lumber had increased 105 percent, while US production had risen only 19 percent. In 1975 Canada shipped 52 percent of its lumber production to the US and, in 1985, 65 percent. In 1975 Canada controlled 18.7 percent of the US market; in 1985, 33.2 percent. Because most of Canada's softwood lumber was produced in BC, this province was the main cause of the problem.[12]

The Coalition blamed the situation on low stumpage rates in Canada. It calculated the average stumpage paid in the US as $20.83 (US) per cubic metre, while Canadian producers paid only $2.37. If Canadian stumpage had been raised $18.46 per cubic metre, as the Coalition suggested, stumpage and royalties on the 73 million cubic metres of Crown timber harvested in 1986 would have amounted to $1.35 billion (US), compared to the $173 million (Can) actually collected. The Americans identified the difference as a subsidy to Canadian lumber producers.

An alternative solution would have been to require Canadian companies to pay a fair market price for timber; that is, the price a willing buyer would pay a willing seller in a competitive market. By comparing stumpage fees on Crown land in Canada with stumpage on private Canadian timber, with BC's SBEP timber sales and on comparative US sales along the international border, the Coalition concluded Canadian lumber, on average, was selling for $13.90 (US) a cubic metre below market value. Other solutions proposed were to establish an import quota on Canadian lumber, or for Canada to levy an export tax or a reforestation tax on all exported lumber. The Coalition dismissed the idea of lifting Canadian restrictions on log exports.

Again the Canadian negotiating team fought the case on legal merits, attempting to prove there was no subsidy. It argued that across-the-board stumpage comparisons were not valid because Canadian loggers incurred higher costs than their American counterparts. As well, it contended, Canadians were far more efficient sawmillers than Americans. Outside the formal legal dispute, a number of other arguments were offered, the chief one being that Canada was free to adopt whatever forest policies it wished and US actions were an unwarranted intrusion into Canadian sovereignty. Canadians who suggested Americans might have some valid arguments were accused of treason. Canadian media, with rare exceptions, took the side of the Canadian lumber industry and almost never reported the US side of the story. It had become a politically charged affair in Canada, which was not surprising considering the Canada–US Free Trade Agreement was under negotiation at the same time. What the numbers actually told us about the application of forest policy in Canada, and BC in particular, got lost in the frenzy to defend Canada against Yankee meddlers in our affairs. Sam Gibbons, a Florida congressman, said during a visit to Vancouver over the issue, "the Canadian government and its officials are convinced that there's no subsidy. I've seen it before—people come to believe their own b.s. if they repeat it often enough."[13] A valuable opportunity to analyze and strengthen forest policy had been lost to Canadian jingoism.

As the case worked its way through the US legal system, it became apparent Canada would not back down from its insistence that there was no subsidy to Canadian loggers, the Coalition would win its case, and a tariff would be imposed. By this time the federal government had taken over negotiations in Canada and its chief negotiator was none other than Pat Carney. She knew that this was a political issue and that at this stage the only question remaining was how high the tariff would be. Her mission was to keep it as low as possible. In October 1986, the US Commerce Department announced its preliminary decision: Canadian lumber was subsidized to the tune of 15 percent and a tariff reflecting that figure should be imposed.

The Canadian team was in disarray; its members began attacking each other publicly and viciously. Leadership of the forest industry's interests had been assumed by Adam Zimmerman, head of the Toronto company, Noranda, whose principals wanted to continue the legal fight. If it was lost, Canadian lumber producers would pay $600 million a year into the US treasury. Zimmerman had accused Carney of appeasing the Americans and said sending politicians to negotiate trade deals was like asking a hospital orderly to perform heart surgery. One of Carney's cabinet colleagues responded that Zimmerman "does not care one whit whether the money goes into American pockets rather than Canadian ones because, either way, this is just another business cost to him. He pays it anyway. He doesn't care to whom." This was not quite true: the industry knew from long experience that it was easier to get a tariff removed than to get stumpage rates lowered. Also, a tariff on Canadian lumber would only apply to volumes shipped to the US, whereas increased stumpage fees would apply to all timber cut in Canada.

BC's new premier, Bill Vander Zalm, settled the issue in late 1986 by offering to raise stumpage rates. This would keep the money in BC, rather than have it end up in Washington if a tariff were imposed. Carney was eventually able to negotiate a 15 percent Canadian export tax on lumber destined for the US. A year later BC raised stumpage fees an equivalent amount and stopped reimbursing logging companies for their reforestation costs.

After signing a Memorandum of Understanding (MOU) in which the terms of the softwood lumber agreement were spelled out, politicians and numerous industry spokesmen produced a great deal of talk about the pact being unjustified, an intrusion on Canadian sovereignty and generally unnecessary. By 1991 BC had substituted an increase in stumpage rates for the export tax, and other provinces had boosted expenditures on forest management, which, they claimed, answered American objections to Canadian policies. That September, the federal government unilaterally terminated the agreement, as allowed for in the MOU.

By this time, the Coalition for Fair Lumber Imports had become one of the best funded and organized lobby groups in the US, and it was quick to apply pressure in Washington. Immediately the US government required Canadian lumber exporters to post bonds on shipments crossing the border and follow other bureaucratic procedures, which pushed the cost of getting lumber across the border higher than it had been under the export tax. BC lumber producers, with increased stumpage charges to pay, were hit particularly hard. A war of words ensued, in which the US industry argued even more strongly against free access to American lumber markets by Canadian producers. Instead, they called for free trade in all forest products, including logs, and raised the issue of the negative impact cheap Canadian stumpage has on US-managed forests. They maintained that low stumpage on Canadian timber cut from natural forests tended to undermine stumpage values in the extensive plantation forests grown on private land, particularly in the southeastern US.

Pat Carney's response was that "each country has its own way of pricing natural resources and they're basically incompatible." She went on to say that "our forests are used to support our communities."

It was an absurd situation. Canadian politicians, who are not always the most capable representatives of Canadian interests, were again trotting out the century-old Canadian position—that we want to cut down our forests and freely export processed forest products to employ our workers, and that it is incumbent upon the US as a trading nation to admit our lumber. Americans point out this is a pretty stupid thing to do, and we should be charging higher stumpage fees so we can reinvest more money to keep our forests healthy, as they do on private land in the US. Canadian politicians respond by saying we have our own unique system—state-owned forests—and if we choose to spend our money on socioeconomic programs instead of renewing forests, it is our sovereign right to do so. Fine, the Americans respond, but if you want to operate that way you may not sell cheap lumber in the US, undermining our sawmill industry and, more important, our managed forests.

It is an issue over which American environmentalists and the forest industry find common cause, and many Americans believe it is morally irresponsible for the US to import cheap forest products from nations like Canada that are not caring for their forests. They feel the US should become more self-sufficient in timber and Americans should pay more for forest products rather than contribute to the deforestation of other countries, including Canada.

In 1992 the US Commerce Department ruled again that Canadian lumber was subsidized and proposed a duty of 14.48 percent on it.

The Canadian and BC governments took the issue to a binational panel set up under the newly minted Canada–US Free Trade Agreement. The panel, with two Canadians and one American on it, ruled against the US in 1993. Jubilation over this victory was short-lived: the following year, the US industry launched a court case alleging that the trade panel violated US sovereignty. As this case made its way through the courts, the US economy took a jump and the demand for lumber shot up. Within a year Canadian lumber shipments had risen from a 27 percent share of the US market to a 35 percent share, touching off a new skirmish with the Coalition.

This time, politicians on both sides of the border handled negotiations, with representatives from industry participating from the sidelines. Canadian industry representatives maintained the position that they enjoyed no subsidies and that apparent differences, such as lower stumpage rates in Canada, could be explained. In the US, the Coalition for Fair Lumber Imports watched the talks and prepared to file another countervail application. Most Canadian media championed Canadian sovereignty, with snide comments about the treachery of shrewd Yankee traders. In February 1996, hours before a deadline imposed by the US after fourteen months of talks, the politicians emerged to announce an agreement.

Canadian and BC politicians were triumphant. BC Forest Minister Andrew Petter crowed about the deal, which had averted a trade war: "Throughout the negotiation process, the full participation and united support from the leaders of BC's forest industry made sure that the BC position was one that would resolve the issue."[14]

Under a five-year agreement that began April 1, 1996, Ontario, Alberta and Quebec agreed to raise stumpage rates and impose stiffer forest management obligations on their industries. Alberta, Ontario and BC also agreed to a quota on lumber shipped into the US, with an escalating export tax imposed after imports reached 14.7 billion board feet a year. As had been the case almost sixty years earlier, the deal was not the triumph its negotiators made out. Within a year, managers of forest companies were realizing what they had actually agreed to in Washington. The first stage of the process involved a nasty fight over how the quota would be divided up among the provinces. Then the federal government set about allocating precisely how much lumber each company would be allowed to export to the US duty-free, setting off another round of bickering over individual quotas. Eventually the most highly regulated forest products industry in the world advanced to an even higher bureaucratic plateau, with full federal regulation of its export lumber sales. As Terence Corcoran wrote in *The Globe and Mail*, it was probably the most "bizarre trade initiative [in Canadian history]—a Canadian tax to make Canadian exports uncompetitive."[15]

The effects of the agreement have been mixed so far. A strong US market during late 1996 and early 1997 raised demand to the point that American customers were willing to absorb the high export taxes when quotas were exceeded. In some cases the deal also encouraged a higher level of manufacturing in Canada, particularly in BC, because lumber products that had undergone more processing were not included in the agreement. However, by the end of 1997 the US market had softened and the Asian market collapsed. With prices down and the Interior lumber industry selling at quota levels or lower, it was impossible for coastal companies, all with little or no US quota, to substitute US buyers for their lost Asian customers. The result of this deeper excursion into forest sector regulation was to lock the industry into an even smaller box than before. The only recourse, in early 1998, was for a joint industry-government delegation to plead with the Americans to permit a reduction in stumpage rates. Also, the US Customs Department began reclassifying various forest products as softwood lumber and, therefore, including them in the quota arrangement.

By mid-1999, with the agreement scheduled to expire in less than two years, both the Canadian and BC industries were deeply divided over what position to take in forthcoming negotiations with the US. The only point on which there was agreement was that the Americans would certainly come to the table with even tougher demands. In BC, coastal and Interior companies were split over the question of quota distribution: coastal producers wanted more and Interior mills, remembering decades of domination by coastal multinationals, said, in effect, "Get lost." Ontario lumber producers, with newly expanded sawmill capacity, wanted more quota—at BC's expense. Provinces outside the quota arrangement, which had increased shipments to the US, were fearful the Americans would insist on their inclusion in the new deal. In short, Canada's position, always weak, was crumbling.

It is not surprising that the log export issue arose during the last lumber dispute. The issue has been part of the Canada–US lumber trade since the day it began. During the 1860s, when Ontario began developing a domestic sawmill industry which saw the US as one of its primary markets, a series of policies were adopted to give a competitive advantage to Ontario producers. Two of these were retention of Crown land ownership and restrictions on log exports. When BC adopted this policy in 1901, the province was drawn into the lumber trade war with the US, and the log export issue has been an inseparable part of BC's timber pricing policies ever since.

During World War II, the federal government became involved in regulating log exports as well, for purposes of national security. When the war ended, the regulations remained in force. They were rarely exercised until 1969 when restrictions were tightened along with those of BC to encourage manufacturing in the province. Today there are two kinds of restrictions. The first requires a permit from the province and the federal government to export all logs except for the small portion cut on private land granted before 1906. An industry-based Log Export Advisory Committee advises the two governments on individual applications, depending on whether the logs are needed by domestic mills. Essentially, if anyone in BC wants the logs an applicant wishes to export, the application is refused. The second means of restricting exports is a log export tax—known as a "timber tax" in deference to the Americans—which serves as a disincentive to export. A similar process applies to pulp chips, the export of which is also restricted. The result is that in most years, log exports represent only 1 to 2 percent of the province's annual timber harvest.

LOG EXPORTS
(in cubic metres)

1966	236,700
1967	339,000
1968	309,000
1969	259,400
1970	500,000
1971	281,100
1972	101,900
1973	47,500
1974	210,000
1975	150,100
1976	589,700
1977	1,033,100
1978	686,400
1979	730,100
1980	1,135,600
1981	903,800
1982	1,275,600
1983	2,308,200
1984	3,322,200
1985	2,400,000
1986	2,303,000
1987	3,890,000
1988	2,529,000
1989	1,565,000
1990	776,000
1991	1,359,000
1992	1,260,000
1993	1,180,000
1994	507,000
1995	764,000
1996	410,000
1997	269,000
1998	1,133,000

Source: Ministry of Forests Annual Reports.

Export levels were higher during the 1980s because of industry's desire and government's willingness, during a period of depressed markets, to raise exports in order to maintain employment in the logging sector; also because of Japan's willingness to pay unusually high prices for hemlock and balsam logs, which were almost unsaleable in BC at that time. This condition repeated itself in early 1998. Most of these exports came from the big forest companies and sparked protests by some of the small independent sawmills who could not find logs to fill lumber orders. The IWA also objected, invoking its long-standing opposition to log exports as a threat to sawmill jobs. A union brief written in 1985 cited a US study that found log exports created 4.72 hours of direct employment for every 1,000 board feet of logs, while milling them into lumber created 12.58 hours of direct employment, and turning them into plywood generated 19.47 hours.[16]

The main argument for restricting log exports is to provide jobs in timber processing plants in BC, and it is successful in performing that task. But there is another, unstated, reason: to provide sawmills with a captive supply of low-priced logs, which gives them an advantage in export markets.

We could approach this issue in a more enlightened way if someone calculated the cost of this policy to log owners, primarily the provincial government. Enough logs are sold outside the country by private landowners to give us an idea of how much revenue-export restrictions are costing us. In 1993, a prominent BC forest landowner thinned two coastal Douglas fir stands, one on land granted before 1906 and freely exportable; the other on land granted after that date and not exportable without permission. The logs in both lots were essentially the same mix of grades and species. The stumpage rates the owner received for the exportable logs averaged $44 a cubic metre; for the non-exportable logs, only $12. Thus the exportable logs were worth about 3.5 times as much to him as the ones he was forced to sell in BC.

Let us assume, for a moment, that these two lots of logs are representative of the entire 1993 BC timber harvest and worth the same amount, and that the entire timber harvest that year could have been sold outside the province, or even inside it, at those prices. Obviously, if that volume of logs were dumped onto the export market in one year it would depress prices. But in time, as buyers gained confidence in the reliability of the supply, mills would be built to cut the logs and prices would be restored.

In 1993, BC mills employed about 70,000 workers. That year, about 70 million cubic metres of timber was cut on Crown land in BC. The average stumpage paid was $14 a cubic metre, providing a total of about $1 billion in stumpage and royalties. If, like the small landowner, the government allowed those logs to be exported and obtained 3.5 times as much stumpage, total revenues for the year would have been $3.5 billion, a gain of $2.5 billion. That amount is what it cost in 1993 to maintain the 70,000 jobs in the province's forest product mills—about $35,000 a job. Of course there were also benefits. A portion of the forgone revenue came back to the government in the form of taxes paid by the millworkers. But even that portion would be partially offset by the taxes that would have been paid by longshoremen and others employed in the log export business.

The hypothetical figures are not perfectly accurate, but they are close enough to show how expensive it is to restrict log exports. And there are other good reasons not to restrict them. For one thing, the major intention of restrictions is to lower the value of timber, which has far-reaching consequences in BC. First, the reduction in value reduces the timber supply. As prices drop, marginal timber costs more to harvest than it is worth on the market and is deleted from the available supply. Second, export restrictions reassign timber value away from the forests where they grow and into the sawmills, pulp and paper mills and plywood plants that process it. The landowner—in most cases, the government—receives less of the value and is therefore unable to reinvest it in maintaining the resource. The $2.5 billion in stumpage value forgone in 1993, for instance, would have paid for all economic silvicultural opportunities several times over. Moreover, by transferring wealth to the mills, large urban centres end up being developed at the expense of rural areas. This process is most obvious on the BC coast, where more than a century of logging and manufacturing has left rural areas relatively devoid of infrastructure.

By suppressing timber values and failing to reinvest at least a portion of that value in the forests, our policies have reduced the timber supply, which in turn has caused mill closures, particularly on the coast. Exacerbating the situation is the availability of underpriced logs, which for many decades has encouraged mills to remain inefficient and workers to be unproductive.

The current condition of the BC forest industry suggests that the long-standing policies on timber pricing and log exports have outlived whatever usefulness they may have had in the past, when the first priority was to develop a timber processing industry. What is urgently needed today are policies that maintain and enhance the timber supply, without detriment to non-timber forest values.

We could easily pay for all the forest management needed to maintain timber and non-timber values if we were to price BC timber at what it is worth on world markets and insure that the increased revenues were available to forest landowners for reinvestment in the forests. The surest way of making that happen would be to allow landowners, state and private, to sell timber freely on world markets. Even if every mill in BC were to shut down—an unlikely scenario—we would still be ahead of the game. In BC we have the resources to build a vibrant, healthy timber farming industry. That industry could provide a reliable supply of high-quality logs that would give an ongoing competitive advantage to an efficient domestic timber processing sector—which would quickly establish itself here if a free and open timber market were in place.

Timber Harvest Rates 8

Among the policies that have the most direct impact on the BC forest industry are those regulating timber harvest rates. The rules on how much timber can be cut each year dictate how many mills will operate and how many workers will be employed. And because these policies modify the structure of forests regionally, they profoundly influence the future of the province's forest economy.

The process of determining how much Crown-owned and Crown-regulated timber can be cut each year is not straightforward. The allowable annual cut (AAC) of each Tree Farm Licence and Timber Supply Area, although in the end a mathematically precise figure, is arrived at by juggling broad estimates of timber volumes and growth rates, social and economic considerations and political pressures. Consequently the process itself and the decisions it leads to are controversial and open to criticism.

The need to regulate the annual timber harvest first arose with the adoption of sustained yield policies in 1947. Before then the amount logged each year in BC was determined by a combination of what markets demanded and what the industry was capable of cutting and processing. With periodic fluctuations, often major ones, the annual harvest had increased steadily since the industry began in the last century.

Contrary to popular misconceptions, the adoption of a sustained yield harvest regulation system was not implemented to restrict the actions of an expanding forest industry about to log itself out of existence. Initially it was intended to create a certain type of forest, which would then produce a predictable and consistent volume of timber annually. Today this volume is known as the Long Run Sustainable Yield (LRSY). This concept was based on the questionable premise that a stable rate of timber harvest could be maintained, and that this would produce a stable economy.

The concept of sustained yield came from Europe, where it was a consequence of the way forestry evolved, rather than a policy objective. As Fred Mulholland reported after studying forestry in Finland and Sweden in 1937, "every forest is managed in the manner which its local forester thinks will produce the best continuous results. In the aggregate this produces a national sustained yield."[1]

In jurisdictions with large areas of private forest land, it is usually not possible for the state to regulate the yield of the entire harvest. Individual owners regulate themselves. In those countries with a developed forest management capability, annual harvests roughly approximate annual growth, over the course of a business cycle. In a country where all the commercial forests are managed forests, there is no need to regulate for sustained yield: it is a consequence of good management.

In 1947, when BC decided to adopt a sustained yield policy and retain state ownership of forests, the Forest Service had to decide how fast to cut existing old-growth forests. While there was widespread agreement on the final objective of a stabilized forest industry whose annual timber harvest was equal to annual forest growth, there were conflicting ideas about how to get there.

At that time the question of whether too much timber was being harvested, even in the heavily logged Vancouver district, was not a pressing issue. A major concern of administrators was that the previous practice of harvesting whatever volume the market would bear was producing a new forest of uneven age classes. Forest theoreticians preferred the idea of a forest in which areas holding equal volumes reached maturity each year, called a "normal" forest in the words of Bernhard Fernow. He believed a "normal" forest could be created by taking a management unit consisting of old-growth forest, such as one of today's Timber Supply Areas or Tree Farm Licences, and dividing it into smaller areas that were equally productive. The number of these smaller areas would equal the rotation in years (the age at which the trees would be deemed "mature") of the new forest. One area would be logged each year, producing a new forest in which a portion reached maturity each year. For instance, if the TSA or TFL contained 100,000 hectares of equally productive forest land, and the planned rotation of the new forest was 100 years, then each year 1,000 hectares would be logged. By the end of the first 100 years, the first cutting block logged would be 100 years old, and ready for a second harvest. This cycle could be repeated indefinitely, providing a steady supply of timber.

Around 1950 the BC Forest Service adopted a more complex version of this rather simplistic system. The Hanzlik formula, named for the US forester who devised it in the 1920s, provided a means of calculating an annual harvest level in a Timber Supply Area containing a mix of old-growth or mature timber, and young immature stands. This kind of system was needed in BC, where most sustained yield units contained such a mixture, as a result of logging or fires. Essentially the Hanzlik system calculated a harvesting volume by dividing the volume of mature timber in the unit by the number of years in the rotation of the future forest and then adding the average annual growth of the immature stands in the unit.

At the time the formula was created, many people felt the Forest Service was becoming unnecessarily concerned about yield and was adopting an overly restrictive system of rules and regulations to control the cut. This opinion surfaced during Sloan's 1955 Royal Commission, by which time annual harvest figures had been established by the Forest Service for each Public Sustained Yield Unit, and approved for each Forest Management Licence where the licensees were responsible for calculating the AAC.

To illustrate its point, the Forest Service presented Sloan with a calculation of the AAC for the Vancouver forest district, obtained by application of the Hanzlik formula. Its figure was 540 million cubic feet. Several professional foresters working for large companies on the coast started with the same figures and the same formula, and came up with an AAC of 654 million cubic feet. The difference reflected the subjective nature of the process. For example, some industry foresters used a lower rotation than the Forest Service's ninety years, based on the assumption that by the time these young forests were harvested, the mills would be able to use smaller logs more efficiently.

However, as Sloan pointed out in his report, the real issue concerning rotation was not the size or quality of the logs to be obtained from the new forest at some distant point in the future, but the immediate impact on old-growth AAC. "Selection of a sixty-year rotation for the immature stands in place of ninety years would increase by 50 percent the annual rate at which the mature timber could be cut."[2]

The question of how fast to liquidate the old-growth forest provoked widespread disagreement. Many people, particularly in the Forest Service, argued that old forests should be logged as slowly as possible, and logs from second-growth forests should be phased in gradually. This way, succeeding generations would be assured of adequate high-quality timber, and annual harvests would not have to be reduced when the old-growth timber ran out. Some observers went so far as to assert that the annual cut in the Vancouver district should be restricted to the volume of annual increment or growth in the second-growth forest.

Others maintained the old-growth forest should be cut as rapidly as possible because so much of it was "stagnant," consisting of "overmature" trees that were not growing. In fact, merchantable timber volumes in many areas were declining. Logging these forests and replacing them with young, new forests that were growing at their full productive capacity, with a proper gradation of age classes, would optimize the sustainable yield much sooner. Logging stagnant forests slowly would mean forgoing economic opportunities today for the sake of future generations, who might have far different timber requirements.

Even C.D. Orchard found himself harbouring such thoughts. In 1952, while attending a meeting of the Western Forestry and Conservation Association, he confided to his diary: "We can't predict wood needs and utilization 100 to 150 years from now. Foresters have been too prone to scare themselves and the country with predictions of famine; and have done too little to forestall their predicted calamity. Sufficient to the day is the evil thereof. Let us (1) get our lands under crop, (2) protect the crop, (3) build and maintain a good inventory, (4) seek to improve utilization, (5) create conditions that will permit forestry, taxes, tenure, etc., (6) let our grandchildren do their own worrying."[3] As Orchard later noted, this was an "inexplicable lapse" on the part of the chief architect of BC's yield regulation system.

The Hanzlik system had other weaknesses. It made no distinction between the volume of timber and its value. The whole concept of a "rotation" was based on the premise that at a certain age the growth rate of a tree falls off. The fact that many of these trees would be adding clear, fine-grained wood worth much more in the market than younger, faster-growing wood did not come into the equation. The system focussed the industry's attention on timber volumes, and reinforced a tendency to ignore the quality of wood.

The system also ignored the reality of global timber markets. Cut control policies were designed to harvest mature timber at a rate that would create the "normal" forest, with its even gradation of age classes; consequently the AAC came to be viewed not as a maximum cut, but as a desirable or even a mandatory cut. Those who did not cut their AAC were eventually threatened with the loss of the uncut portions. If the timber was not harvested at the indicated rate, the argument went, the age–class gradation would be distorted and at some point in the future would lead to either a shortage or a surplus of timber.

This policy might have made sense had the forest products manufactured in BC been sold into an insatiable market. But that has never been the case. BC lumber, plywood, pulp and paper are sold into international markets that fluctuate because of factors beyond BC's control. The system, with a modicum of built-in leeway, forces industry to maintain production during sagging markets, driving prices even lower and selling off the province's timber inventory at bargain basement rates. At the other end of the market cycle, when prices are high, cut-control policies restrict the amount of wood that can be processed and sold, and BC loses potential increases in public and private revenues. The impact of these policies on markets should not be underestimated; as noted earlier, BC provides 35 percent of the world's softwood lumber exports. Ironically, the major moderating influence on this condition is the small amount of private timber that is managed outside the yield regulation system, whose owners sell when prices are high and let their trees grow when they are low.

Another weakness in the system is its creation of an "allowable cut effect" (ACE). The Hanzlik formula and its variants dictate that any change in the growth rates of immature forest in the timber supply unit affects the volume of mature timber that can be harvested each year. Silvicultural treatments that improve growth rates, or are perceived to improve growth rates, drive up the AAC, which stays higher over the rotation of the treated forest. The rate of financial return on investments in such silvicultural treatments can appear very high when the value of the AAC increase is included in the calculation. Many of the silvicultural programs initiated by industry and government over the last thirty years relied on ACE for their acceptance. As a result we have very little idea of the real returns on silvicultural investment, much of which had little positive effect on the forest, but yielded large returns in the form of a higher AAC.

ACE also distorts the impact of events such as fires or infestations that impede the growth of immature forests. Because it spreads the timber loss over many years or decades, it reduces the apparent value of protecting these forests.

ACE works in many other ways. Acquiring logged-off land and adding it to the sustained yield unit enlarges the stock of immature growth, which produces an immediate and sustained increase in AAC, which may be worth far more than the cost of the additional land itself. This factor came into play when old temporary tenures were logged and then rolled into Tree Farm Licences under provisions in the 1978 Forest Act amendments.

In his 1976 Royal Commission report, Peter Pearse described weaknesses inherent in the system and questioned many of the assumptions on which it is based. Like Sloan, he took issue with the argument that "future generations will be best served by the balance of old-growth and new forests that will result from the current allowable cut limits." He went on to say:

> It is a dangerous oversimplification to assume that we can dispatch our obligations to the future by harvesting at a constant rate, or even that the result would be preferable to the forest structure that would emerge from some other regime.
>
> Another spurious justification for steady harvesting is that it protects non-timber values in the form of watershed control, fish and wildlife, and recreation. But the rate of harvesting is much less important to the protection of the forest environment than choices about which timber is to be removed and which preserved, the logging and road building techniques to be used, the pattern of clear-cut openings, and post-logging treatment of the site.[4]

Pearse also challenged the notion that BC's yield regulation policies are essential to the maintenance of regional economic stability because they help moderate harvesting and milling rates under the influence of market fluctuations. Application of the policies aggravate short-term market instabilities, he reported, and do little to prevent long-term destabilization. He concluded:

> It cannot reasonably be assumed that the present controls will result in a steady yield of timber, in spite of their implied objective. Repeated revisions of allowable annual cuts in the past testify to the transient relevance of long-term predictions based on assumptions of a constant technological and economic environment, and continuing change can be expected in the future. Moreover, even if the timber supply were constant, it does not follow that industrial activity and the com-

> munities based on it would remain stable. In some cases, undoubtedly, the survival of communities warrants action on the part of governments, but history suggests that this calls for measures other than the regulation of harvest rates.[5]

Pearse's major recommendation for reform consisted of a fundamental shift in the objectives of yield regulation policies.

> It is no longer adequate to fix the goal of yield regulation as a more or less constant flow of timber volume over very long periods. Present circumstances call for a more flexible approach to yield regulation, with greater emphasis on protecting and enhancing the productivity of forest land and on the economic, social and environmental implications of harvesting. Forest management policy for the future should be directed toward two related objectives: protection and enhancement of the capacity of forests to produce their potential range of industrial and environmental values; and within that framework the regulation of harvesting to produce the maximum long-term economic and social benefits from the timber resource.[6]

The implications of such a shift were considerable. Forest land should be put to its best use, which might not be the production of commercial timber; timber harvesting should be planned to take advantage of markets and technological trends, rather than to produce a "normal" forest; reforestation should be done promptly and silvicultural investments made on the basis of the benefits they provide for the new forests, rather than as a means of raising the AAC.

By the time Pearse's recommendations had worked their way through the bureaucratic mill, the results were somewhat different than those he sought. The Ministry of Forests adopted his suggestion to redraw the sustained yield units, with a reorganization of the eighty-eight existing PSYUs into thirty-three Timber Supply Areas (TSAs). Computer modelling programs were used to massage the available inventory data, much of it of questionable accuracy, to estimate the Long Run Sustainable Yield (LRSY) of each TSA. When these figures were published in 1979 they set off alarms throughout the industry and beyond about an impending timber supply "falldown." This term described the discrepancy between the annual harvest of mostly mature timber permitted under the cut control system, and the amount of timber the new managed forests would produce, the LRSY. In most TSAs and TFLs, particularly on the coast, the LRSY was smaller than the current AAC, prompting a two-pronged flurry of activity. The Forest Service immediately called for reductions in AAC, which were lowered to levels already committed to industry under leases and licences. As well, a cry went up about the need to spend more on forest management to maintain the timber supply. With a great deal of prodding from industry and the Forest Service, provincial and federal governments were persuaded jointly to fund a variety of silvicultural projects, including replanting the backlog of inadequately regenerated lands, spacing juvenile stands and fertilizing. The largest of these programs, the 1984 Forest Resource Development Agreement (FRDA), directed $300 million at these projects, and in 1991 FRDA II committed another $200 million to them.

The results were somewhat contradictory. In spite of concern about cut levels, the provincial AAC rose from 62 million cubic metres in 1976 to 67 million cubic metres by 1980, and to 71 million cubic metres by 1992. A confluence of factors had pushed the AAC steadily upward.

One factor was the Bill Bennett Social Credit government, which held office throughout most of this period, and which was more sympathetic to the short-term financial interests of big timber companies than almost any administration in the province's history. This government was very amenable to maintaining or raising the AAC over the Forest Service's objections. In 1983 Bennett announced a budgetary "restraint" program and reduced the size of the Forest Service by 30 percent, undermining its ability to gather or analyze the inventory data needed to recalculate AAC. The following year a policy of "sympathetic administration" was adopted, under which forest management and logging regulations were selectively ignored or relaxed. Large amounts of low-value, merchantable timber were logged to roadside and burned, without being included in AAC calculations.

A modified method of determining the AAC was adopted. It incorporated the ACE of new forest management programs, a rejigging of TSA boundaries and numerous other factors, resulting in higher cut figures. Significant political pressures were exerted on chief foresters to ensure cut levels were not reduced.

At the same time, downward pressures on the AAC were developing. In order to protect other forest values, various restrictions on logging were being introduced, ranging from outright prohibitions to special procedures. Many of these restrictions should have been accounted for in cut control reviews, and should have resulted in reduced AACs, but these calculations were rarely made during the 1980s. In fact, the operations division of the Ministry of Forests took the position that no such restrictions on logging could be imposed and no other values acknowledged. Commitments to forest companies, no matter how unrealistic, must be fulfilled.

In April 1991, only a few days before the Forest Resources Commission report was released, Chief Forester John Cuthbert sent a pointed memo to the Ministry of Forests chief of operations, Wes Cheston, instructing him to abandon this line of argument, and ordering that "we stop further erosion of our credibility by taking rigid stances in our planning processes regarding the implications of AAC or existing land base assumptions."[7]

By the time the Forest Resources Commission was established in 1989, a widespread debate over harvest levels had developed. Many felt that the yield regulation system was out of control and that cuts were too high, because in some TSAs and TFLs they were substantially greater than the LRSY. Others argued the cuts were not too high, and that some should be raised to liquidate slow-growing old-growth stands and replace them with healthy, productive new forests as quickly as possible.

The debate, which was similar to the debate of the 1950s, reflected opposing responses to the falldown. That a falldown will occur in most sustained yield units is not disputed. A new managed forest that replaces an old-growth forest will not, during the rotation chosen for the new forest, produce as much merchantable timber as the old-growth forest on that site provided. One argument is we should avoid falldown by establishing the AAC at the same level as the productive capability of the site, forgoing economic benefits today for the sake of a stable economic future. The other argument is that we should not sacrifice a current economic opportunity, nor should we do without the more rapid growth the young forest would be accumulating if the old growth were harvested and replaced, without any certainty that future generations will want or need the kind of wood found in old-growth forests.

Since the AAC was higher than the LRSY in most of the province's sustained yield units, the debate focussed on when the two numbers should be brought into alignment. Should we grapple with the consequences of the falldown now, or put it off and cope with its consequences later, over a shorter span of time?

Like many forest policy debates, this one was not simple or straightforward. Environmentalists and others pressing for logging restrictions often invoked the falldown argument in support of their own arguments for restricting or prohibiting logging. Numerous submissions to the FRC urged big reductions in the harvest of old-growth timber and big increases in silviculture spending to improve growth rates in young forests and avert falldown.

The FRC concluded the issue could not be resolved then because "the current state of inventory information is a disgrace,"[8] and that the AAC of many Timber Supply Areas could not be determined with confidence. The FRC report went on:

> The current timber endowment consists largely of "old growth" trees that have a greater volume of wood at harvest than will the "second growth" trees that replace them, under current management regimes. Those current timber management regimes and yield calculations are not designed to replace old growth inventories with an equal volume of wood.

> That means there could be a "falldown" of the inventory volume. This does not necessarily imply that current harvest levels are not sustainable, but it has led to a perception that BC's forests are being overcut. It also doesn't take into account the possibility that through intensive forest management the timber volume could actually be increased in some areas.
>
> The real problem . . . is that the information used to calculate these yield results is very unreliable. Yield analysis is often completed with an inconsistently defined estimate of the working forest area, outdated and unreliable inventory estimates of volume and production capacity, and growth and yield forecasts that do not reflect the actual growth rate of the forest. Given the uncertainties in the data base . . . there can be only limited confidence in the yield analysis results. In other words it is possible that the AAC determination is too high. It is equally possible that the AAC determination is too low.[9]

The FRC recommended the AAC of most TSAs and TFLs be frozen until new timber inventory data and a new yield analysis were available, except in the few areas where the current information made it obvious the AAC should be adjusted.

By this time the Ministry of Forests had undertaken an internal review of the timber supply and the chief forester had concluded, for various reasons, that the AAC of some TSAs and TFLs needed to be lowered. Restrictions on logging to allow for other resource values had not been incorporated into estimates of the available timber supply, the anticipated gains in growth from the expensive silvicultural programs had not occurred, and many inaccessible timber stands, which industry had said it would be able to log with new methods, were still not economically harvestable.

Both the Bennett government and the Bill Vander Zalm government that replaced it in 1986 refused to go along with the Ministry of Forests' recommendation to undertake a major timber supply review, with adjustments in the AAC to follow. This situation changed after the NDP was elected in 1991 and the chief forester was instructed, in an amendment to the Forest Act, to complete a reassessment of the AAC by the end of 1995, and to determine a new figure every five years for every TSA and TFL in the province. NDP Forest Minister Dan Miller stated unequivocally he did not intend to interfere in his bureaucrats' work.

In 1992 the Forest Service initiated the Timber Supply Review Project, an accelerated three-year project to reset TSA harvest levels. New levels would be based on available inventory data, which the FRC had deemed "inconsistent at best, and woefully inadequate at worst." A concurrent review of TFL cut levels would utilize licensees' data, generally considered to be more reliable.

The TFL part of the review was complicated by the fact that a year earlier the chief forester had attempted to reduce the AAC in MacMillan Bloedel's TFL #44 by 280,000 cubic metres. The company successfully appealed the decision and challenged the Ministry of Forests' right to impose such a reduction. In response the government amended the Act to affirm the ministry's authority, the AAC was subsequently reduced, and reviews of all TFLs were conducted.

The process by which the AAC was established by the late 1990s has evolved considerably since the Hanzlik formula became the basis of the yield regulation system put in place at mid-century. As the Forest Service makes clear, the AAC "is set through a synthesis of information and analysis and, as such, is a choice or judgement."[10] It is a four-step process, conducted primarily by Forest Service district staff.

The first step is to determine the timber supply of the TSA. Available inventory data, with all its inadequacies, is combined with the latest information on what areas are available for logging, the expected growth rates of the new forest, the consequences of meeting the needs of other forest resource users and values, and anticipated losses due to fire and pests. With the help of computer programs, the ministry uses this information to project the theoretical timber supply over the next 200 years. One Timber Supply Analysis Report is developed for each TSA and these analyses are circulated publicly for review and comment.

The second step consists of a socioeconomic analysis, which estimates the direct and indirect impacts on employment of various timber supply projections. This report is also published.

The third step entails several rounds of meetings with the public, including interest groups, forest workers, tenure holders, local government officials, Natives and just about anyone else who wants to have a say. The Forest Service staff prepares a report summarizing public input and passes it to the chief forester along with the timber supply and socioeconomic analyses.

The chief forester is required by the Forest Act to take into account a wide range of other factors, including "any other information that, in his opinion, relates to the capability of the area to produce timber."[11] He is to consider the factors within a particular TSA, as well as the social and economic objectives of the government of the day.

The chief forester then decides what the AAC will be. Under legislation passed in 1991, he is required to repeat the whole process every five years. It is the toughest and, in a way, the most absurd task the chief forester is required to perform.

Results from the TSA review process began to appear by late 1992, sending shock waves through the industry. In the first three TSAs examined, the process led to AAC reductions averaging 25 percent, and indicated province-wide reductions in the range of 20 to 30 percent. This would translate into a loss of between 15 million and 20 million cubic metres of timber, costing 15,000 to 20,000 direct jobs and double that in indirect employment. As this realization sank in, along with the knowledge that even more cuts would be caused by the designation of additional parks and protected areas, as well as by the government's other forest sector initiatives, consternation and alarm spread through the industry and the province's timber-dependent communities. Environmentalists were elated; they saw the reports as a justification to reduce logging in old-growth forests throughout the province.

During the early stages of its latest timber supply review, the Ministry of Forests appeared to be taking a conservative and long-overdue reappraisal of harvesting levels, tending toward an early, less drastic accommodation of the falldown. In a few TSAs, falldown time had already arrived, and the day of reckoning was coming closer. In the Sunshine Coast TSA, for example, the 1.4-million-cubic-metre AAC set in 1981 was reduced to 1.1 million cubic metres in 1993. The next timber supply analysis in 1995 indicated that this cut level "cannot be maintained in the long term based upon current management practices. The report does project that the allowable annual cut may be maintained for one decade, followed by a decline of 10 percent per cent over 20 years to 876,000 cubic metres per year. It may then be maintained at this level for 100 years before rising to the long-term sustainable timber supply level of 985,800 cubic metres."[12]

In the end, in 1996, the chief forester decided the Sunshine Coast AAC would be raised by 140,000 cubic metres by adding previously ignored deciduous species to the calculations.

In the Mid-Coast TSA, the 1990 AAC of 1.5 million cubic metres was 74 percent above the assigned LRSY of 874,000 cubic metres. In large part this difference was due to exceptionally high visual-quality constraints imposed to accommodate Alaskan cruise boat operators. In 1993, the AAC was dropped to 1 million cubic metres, a figure maintained after the latest review, in spite of the Forest Service judgement "that over the next 50 years the timber supply is projected to decline to 550,000 cubic metres per year if current management practices continue."[13]

The largest coastal reduction was in the Kingcome TSA, where the timber supply analysis indicated reductions of 35 percent were needed, but "a combination of factors, including less constraining visual-quality objectives and the adoption of an alternative transition to the long-term harvest level, allowed a smaller reduction of 25 percent."[14]

In response to growing protests against the level of AAC reductions in the early stages of the TSR process, the Harcourt government implemented two measures. Working through the industry-dominated Forest Sector Strategy Committee, it cobbled together Forest Renewal BC to provide employment for loggers and millworkers who lost their jobs because of AAC reductions. Industry would be funding the plan through increased stumpage fees, but it was persuaded to co-operate because of an understanding that any silvicultural projects funded under the plan would lead to increases in the AAC through the allowable cut effect (ACE). Whether these projects actually improve forest growth remains to be seen: as the Forest Service noted in reference to similar FRDA-funded silvicultural projects, "the anticipated gains from intensive silviculture had not materialized."[15]

The second measure was to postpone until the end of 1996 the deadline for the Forest Service's completion of the Timber Supply Review Project, which was providing the information needed for the new AAC determinations. The postponement delayed the announcement of possible AAC reductions until after the 1996 provincial election.

To complicate the issue even further, questions arose about the validity of the data on which inventory reviews were based. In the Fraser TSA the review process led to an AAC reduction of 12.2 percent in 1995, but the accuracy of the data was questioned. An independent examination revealed that the existing data, obtained in the 1960s and 1970s, assumed an average of 584 cubic metres of timber per hectare, while in reality there was only 452 cubic metres—an overestimate of 23 percent.

In another surprise, toward the end of the review process, results of the Forest Service's Old Growth Site Index (OGSI) survey began to come in. They indicated second-growth forests are growing much more rapidly than had been assumed when the Timber Supply Review was done. If true, these results undermine the credibility of the review process and the consequent changes in AAC. Cut levels can be adjusted upward to reflect these findings in five years, but this is small consolation to forest workers who lost their jobs as a result of the review. There are also some concerns about the OGSI methodology, which may undermine the credibility of its findings.

In the Williams Lake TSA, the AAC was reduced 4.2 percent in the 1995 review. The Cariboo–Chilcotin Conservation Council petitioned the BC Supreme Court for a judicial review of Chief Forester Larry Pedersen's decision. The court found that Pedersen had not accounted for the creation of a new park that removed 2.5 percent of the commercial timber land base from the TSA, and instructed him to clarify and reconsider his decision. He did so, and made no changes in the AAC, explaining: "This park land represents a very small portion of the TSA and most of this land is in highly constrained management zones where only limited harvesting can occur. As a result, removing this land had no impact on the short-term harvest level and only a negligible impact in the long term."[16]

These events, and others like them, raise serious doubts about BC's yield regulation system. The truth is hard to discern: the forest industry naturally thinks some AAC levels are unnecessarily low. Environmentalists and other critics of industry claim they are too high. They and others have also claimed that AAC determinations made under Premier Glen Clark's administration were consistently higher than those made under the Harcourt government. They may be right. The chief forester is required to consider government policies in his deliberations, but we don't know what effect these policy pronouncements have because the government is not required to make public its policies in this matter. As Dave Zirnhelt, Clark's forest minister, explained in another context, "The government can do whatever it wants."

The review and adjustment of AACs in all TSAs and TFLs was completed by the end of 1996 (see Appendix, Table 8A). These figures show that although there was a negligible change in the provincial AAC, some areas and some licensees experienced substantial reductions. Interfor lost 45 percent of its cut in TFL #54 by the creation of protected areas and other restrictions in Clayoquot Sound. West Fraser lost more than 23 percent of its cut in TFL #41, due to the creation of a park in the Kitlope Valley. Cut levels in other areas, particularly in northern BC, were substantially increased, largely through the addition of deciduous species to the commercial timber base. Much of the controversy surrounding the issue of AAC changes is raised by the fact that because of the determination process, it is virtually impossible to know what proportion of the reductions is due to the creation of protected areas and how much by other factors. The Forest Service did not calculate the amount of AAC included in a new park and then deduct it from the AAC of the appropriate TSA or TFL. Instead it simply calculated the new AAC of the industrial forest land left after the park was created. In some areas, such as those mentioned, where large protected areas were created, there are obvious correlations between new parks and reduced AACs. But it is not possible to know with any degree of certainty what the economic impact of new parks has been.

By law the process must be repeated every five years. As the chart in Table 1A (Appendix) indicates, if the government carries out its pledge to protect 12 percent of the province, another 1.5 million hectares is yet to become parkland, some proportion of which will be productive forest land. The government has greatly increased the budget for forest inventory work to obtain more up-to-date data, but if the past review is any indication, the major factor in the process is government policy of the day, and the end result depends primarily upon which way the political winds are blowing at the time.

Considering the crucial functions that yield regulation policies and practices are expected to perform, there is little wonder that they generate so much controversy and, no matter how often they are revised, appear to create as many problems as they solve. As if determining harvest levels were not difficult enough, the system has had numerous side effects. For example, the former Forest Service operations manager tried to use maintenance of the AAC as an immutable law which prohibited protection of non-timber values. Other problems have been more complex. Limitations in application of the cut control system have created a major obstacle to selective harvesting of merchantable timber from immature forests. This type of harvest, usually called commercial thinning, can provide many benefits, silviculturally and economically. It can improve the value of the forest by removing diseased or malformed trees and encouraging only the best and most valuable to grow, it can improve species and stand diversity, and it can even enhance growth rates. It can also provide a supply of timber that covers all or some of the cost of the silviculture, as well as supplying timber to mills. In some European countries, most of the timber supply is obtained through this type of harvest.

Until now, incorporating commercial thinning operations into timber supply analyses presented real problems. The Ministry of Forests put strict limits on commercial thinning opportunities, even though there are hundreds of thousands, if not millions, of hectares of second-growth forests in the province that would benefit from thinning and substantial economic benefits to be gained. The reasons for the restrictions are primarily bureaucratic; a system of rules and regulations adopted for one purpose prevents the adoption of other, desirable measures. The Forest Service has insisted that all commercial thinning volumes obtained must be included in AAC tabulations. Since thinning is more costly than other types of harvesting, and licensees are not permitted to take these cost factors into account, they have demonstrated an overwhelming preference to harvest old-growth timber by conventional means.

At various times in the past, the Forest Service has announced its intention to permit increased commercial thinning, as it did again in its 1994 *Forest, Range & Recreation Resource Analysis*, which stated: "The ministry projects that in 10 years, up to 10 percent of the harvest volume from Crown land in BC will come from commercial thinnings."[17] By mid-1999, there was no sign of any increased thinning activity.

A more serious problem is created by the yield regulation system in respect to tenure. A large portion of the 15 million hectares of second-growth forests in the province's Timber Supply Areas would benefit from silvicultural treatments, which could be completely or partially paid for by commercial thinning and other stand improvement cuttings. These areas could be allocated to Woodlot Licences or other new forms of area-based forest-management tenures, with the potential for enormous silvicultural and economic benefits. The primary obstacle to such an undertaking, according to the Ministry of Forests, is that there is no unallocated AAC in the TSAs to put into these tenures, so they cannot be created.

It is difficult to determine whether the bureaucrats are describing what they genuinely believe to be an obstacle, or if they are merely using this argument as an excuse to restrict the use of tenure forms they are predisposed against. Woodlot Licences have been restricted to about one-half of 1 percent of the provincial AAC, although the NDP government, like the Social Credit administration before it, promised to double the number of licences. Whatever the reason, the existing yield regulation system impedes the creation of forest management tenures in the TSAs which could fulfill the ministry's own stated silvicultural and economic objectives.

One of the characteristics of the yield regulation system first employed in 1947 is that it is used to regulate harvesting activity in both old-growth and new forests. That practice came from the desire to develop a rigidly stratified "normal" forest (stands evenly distributed among a range of age classes), which itself was an attempt to maximize the Long Run Sustainable Yield of the new forest. In BC, with our propensity for regulation and central planning, the Forest Service has attempted to achieve this goal by establishing large sustained yield units, then calculating the volume of wood the government wanted to obtain from them, and devising a convoluted bureaucracy to control the behaviour of forest managers with the aim of maximizing a sustainable yield. The vision of a steady flow of timber running through the mills and generating a constant supply of economic benefits has become the Holy Grail of chief foresters. So far, they have not had much success in their quest.

With a different yield regulation system it would be a relatively simple matter to continue logging old-growth timber at one rate, while establishing forest stewardship tenures of one sort or another in second-growth forests. Timber harvesting activities in the latter would be determined by silvicultural criteria and yield would be limited to an amount less than growth. In time, with every forest manager striving to produce the best possible results, this would, as Mulholland wrote sixty years ago about Finland and Sweden, produce a provincial sustained yield.

As Peter Pearse demonstrated in his 1976 report, the yield regulation system used in BC has failed to achieve most of its stated objectives. It has not created a new forest with an even distribution of age classes, it has not provided either short- or long-term economic stability, and it has not served as a widely accepted means of resolving the contentious issue of how fast to harvest the forests. In the end, the system depends on the judgement of the chief forester, whose thankless task is heavily influenced by his relationship with the government of the day. The side effects of the system provide ample reason to question its effectiveness and sufficient reason to adopt a new method.

9 Silviculture

Silviculture, the art and science of growing trees, has been practised for centuries in various parts of the world. European silviculture has its origins in ancient Greek and Roman times with the development of knowledge and skills in growing certain valuable trees.[1]

Today, with widespread use of tree plantations, it is important to distinguish between silviculture and forestry, which is the managing of communities of plants—including trees—and animals that make up a forest. Before the practice of forestry could develop, a social and legal awareness of the forest had to evolve, as well as a concept of ownership. Until 1,000 years ago, no one claimed ownership of the forest. Forests provided food such as game, nuts and mushrooms, and little use was made of timber until sawmilling developed. Scant attention was paid to forest conservation.

Around the tenth century, in response to changing economic circumstances, a new social system—feudalism—established itself in Europe. One of its tenets was a hierarchy of property rights. Put simply, land within a country or kingdom, including forest land, was owned by the king. He would grant certain rights over some of this land, known as a fief or a seigneury, to a powerful local figure in exchange for cash, goods, the provision of troops during wartime, and other considerations. This person might in turn make similar arrangements with others in his fief. Under French colonial rule, seigneuries were granted in Quebec, providing the basis of private forest ownership along the St. Lawrence River.

As populations grew and technology advanced, forests were gradually displaced or destroyed. By the end of the Middle Ages, new needs for timber began to affect forests. Vast amounts of fuel wood were converted to charcoal for use in smelting metals and providing heat in industrial processes. When populations began to concentrate in cities, forests were cleared to make way for agriculture, and the advent of global exploration also led to a huge new demand for timber to build ships.

In this changing social and economic climate, the ability to grow timber acquired commercial value. Silvicultural skills, which had been evolving since classical times, were in demand and new skills were being developed. For example, by the time of Christ it was known that when some species of trees were cut down, their stumps would send out fast-growing sprouts. This silvicultural technique, known as the coppice method of regeneration, was refined and used extensively to produce fuel wood during the Industrial Revolution.

In this climate, as long as everyone had the right to cut trees in a forest, no one would invest time or money in planting more. But when owners established secure, long-lasting property rights over land or timber, they began to invest in future forests.

During this same time an understanding of forest ecology began to develop. With rapid depletion of forests in the early stages of the Industrial Revolution came an awareness of the role forests play in regulating the flow of water. As headwaters of watersheds were cleared of their forests, for example, more floods occurred in the settled valleys and lowlands. It was observed that certain valuable species of plants and animals disappeared when forests were removed. Very slowly, the value of forest habitat was recognized. By the fourteenth or fifteenth century, some countries began to acquire an extensive forest management capability which included advanced silvicultural skills. A steady demand for forest products, property laws that permitted or encouraged investment in forests, and forest culture, made up of an accumulation of skills, knowledge and social attitudes, made such management possible.

By 1500, British foresters had sophisticated techniques for growing specialized shipbuilding timbers. They knew how to place oak trees so that they grew in the long, sweeping curves required for keels, and how to place rocks among tree roots to produce strong "knees" of the right shape for bracing the internal structure of a ship. Foresters in some of the Baltic states learned how to grow tall, straight coniferous trees with little taper by planting them in dense stands and forcing their growth upward. The timber from these forests commanded premium prices throughout Europe for use as masts and spars.

Ship construction used enormous volumes of the best quality timbers. During the age of exploration and colonization, until the widespread use of steel ships in the late nineteenth century, a country's power depended on the strength of its navy, and shipbuilding timber was one of the most strategically important materials in the world. These same circumstances had prevailed in the eastern Mediterranean centuries earlier, when successive waves of imperial expansion were bolstered by access to the excellent shipbuilding wood found in Lebanese forests. Unfortunately, the ability to manage, sustain and replace them never evolved and these forests were mostly destroyed.

In Japan, which is extensively forested, forest management has been under way for centuries. Some 1,700 years ago the Japanese understood the connection between forests and flooding, and passed laws to regulate the use of forests. Many centuries of hit-and-miss forest use followed, and in 1603 the Tokugawa shogunate ushered in the modern era of Japanese forestry. Silvicultural skills were developed and a large body of silvicultural literature began to accumulate. Even after several centuries of social and political upheaval, Japan's commercial forests have survived and its forest management skills have evolved. Today, with less than one-half of BC's land area and forty-five times its population, Japan enjoys a forest base that covers almost 70 percent of the country. Slightly more than half of it is a managed natural forest that has been sustained over the centuries by some of the world's most sophisticated silviculturalists. Most of the remaining forest consists of plantations in which primarily coniferous species are grown. Ironically, the greatest threat to Japanese silviculture has occurred since World War II, with substantial imports of low-priced logs and lumber—mostly from North America.

Global forest history demonstrates that development of forestry skills is not inevitable in a country blessed with extensive forests. The permanent deforestation of the Mediterranean nations and the fate of the Himalayan forests attest to this. Nor is a country's ability to regulate use of its forests and prevent their destruction a measure of its ability to develop silvicultural skills. In India, for example, a well-entrenched set of forest laws has prevented the forests from disappearing altogether, but the ability to grow forests has evolved slowly.

British Columbia, too, has been blessed with one of the world's great natural forests, an endowment that by no means guarantees a future flow of wealth. The notion that forests can be perpetuated and enhanced to maximize the benefits we obtain from them is still an unfulfilled promise. Forest management has been a pivotal point in provincial forest policy for at least fifty years, and it is the key to a sustainable working forest. Many British Columbians realize that if we managed our forests as well as other countries do, we could maintain and even increase our annual timber harvests in a manner that would enhance non-timber values as well. Organizations with such diverse purposes as the Council of Forest Industries and the Western Canada Wilderness Committee argue that better forest management will resolve many of the problems and conflicts now plaguing the forest sector.

Good silviculture begins with the understanding that forests are diverse and dynamic. Each of the world's forests is unique because a forest is comprised of diverse elements that never combine twice in precisely the same manner. As well, these elements are in a state of constant change. The forest on a particular site today will not be there tomorrow or a hundred or a thousand years from now. It will always be changing, with or without human intervention. Forests cannot be preserved in a fixed state, only as dynamic entities. Human beings are as diverse as the forests they utilize—we all want different things from the forests. Our values are also dynamic. What is necessary today may seem like folly a century from now when the time has come to reap the intended benefits.

Forestry is an act of faith in the continuity of human affairs, as well as in the forester's symbiotic relationship with the forest. Flexibility, responsiveness and a vibrant curiosity may be the forester's most important qualities.

In BC, until about sixty years ago, forest management was just an idea. Valuable stands of timber got some protection from fire but otherwise forests were left to fend for themselves, and whatever was of value was removed. After they were logged or burned, some grew back into healthy, diverse forests with saleable timber and healthy populations of the other species that make up a forest. Others did not. Almost all of them, if some of the silvicultural skills developed in other parts of the world had been applied, would be far more productive forests, economically and ecologically, than they are today.

The practice of silviculture began in BC in the 1930s. Initially this endeavour was limited to encouraging reforestation of logged-off lands. It included after-the-fact activities such as planting seedlings on lands clearcut or burned, and providing for regeneration of trees by modifying harvesting methods. Timber harvesting is now entrenched legally in the silvicultural regime. The classification of silvicultural systems is tied to timber harvesting methods and tree regeneration processes employed as part of forest management.

BC forests are managed through two basic silvicultural systems: even-aged and uneven-aged. Even-aged silviculture employs a repetitive cycle of clearcut harvesting an entire stand, sometimes in a series of cuts. In uneven-aged silviculture, individual trees or groups of trees in a stand are harvested selectively and the area is never cleared entirely. There are many variations on both systems, and the two methods sometimes overlap. Controversy continues over the relative merits of clearcut versus selective harvesting systems.

Most logging in BC has been by clearcut—about 90 percent of the areas harvested since the 1920s. When logging first began in about

1860, it was done selectively. Individual trees, mostly Douglas fir and some western red cedar, were felled, bucked and skidded to the mill or the nearest water with oxen. Smaller fir and cedar, along with such commercially undesirable species as western hemlock, red alder, broadleaf maple and Sitka spruce, were left standing. This method of selection logging prevailed on the coast for the next fifty or sixty years. It was done not out of any concern for the future state of the forest but to avoid the cost and trouble of falling unsaleable trees.

Around the turn of the century, mechanized logging equipment was introduced. Steam donkeys were used to drag logs along the ground to railways, where they were loaded and hauled to the mill or a log dump on the water. Compared to a huge modern clearcut, this type of logging seems benign and environmentally friendly. But like the oxen and horse logging that preceded it, this method cannot properly be considered part of a silvicultural system—it was not part of a permanent cycle of forest use and regrowth, intended or unintended. Rather, it was a form of "high grading," in which the most valuable timber was removed and the remaining forest left to chance. Many of the forests that evolved in the wake of this kind of logging were diseased or otherwise compromised. On the coast, where it was common to remove only the fir, surviving forests were often dominated by hemlock, much of it infected with parasitic dwarf mistletoe. The remaining stands were typically too thick to permit the natural regeneration of fir, which requires exposure to the sun. Instead, the shade-tolerant hemlock established a new generation underneath the spurned, unhealthy veterans. The mistletoe infections were passed on to the seedlings, creating vast areas of badly diseased hemlock stands on some of the most productive growing sites of the coastal forest. Much of the second-growth timber obtained from these sites over the past decade or two is useful only as pulp. As well, those early loggers left large amounts of slash that became fuel for fires that regularly ravaged the picked-over stands.

It is this history of high-grade logging that makes many of today's professional foresters wary of selective logging. They know there is often a fine line between the two methods—a line that changes over time and from one part of the forest to another. At its best, selective logging improves the economic value of the stand; at its worst it degrades economic worth, though the ecological health of the stand may be sustained.

The evolution of high-lead yarding systems, which began about 1920, ushered in an era of progressive clearcutting. Entire valleys were logged, and the wood- or coal-burning machines started fires that swept through the slash and up the mountainsides into the standing timber left behind. It was the most destructive era in the province's logging history, not because it employed clearcutting but because it did so indiscriminately. There was little or no thought to future forest values, including timber. From this point on, clearcutting was the prevalent logging method in BC. With the adoption of sustained yield policies and the beginnings of forest management in the late 1940s, silviculture consisted of even-aged management and planned periodic clearcut harvesting. Apart from some areas in the southern Interior of the province, where selection or uneven-aged silvicultural systems have been used occasionally, clearcut logging and the silvicultural systems that go along with it have prevailed.

Today, according to the Ministry of Forests, approximately 90 percent of the 200,000 hectares that are logged each year are clearcut. Most selective harvesting is done in the hotter and drier southern Interior. There, some form of partial cutting is necessary because shade is required to re-establish any forest.

SILVICULTURAL SYSTEMS 1996–97

(hectares harvested)

Region	Clearcut	Partial Cut	Total
Cariboo	36,382	7,641	44,024
Kamloops	19,950	10,044	29,994
Nelson	15,712	7,507	23,219
Prince George	46,497	2,282	48,779
Prince Rupert	20,757	1,728	22,485
Vancouver	28,621	1,908	30,528
Totals	167,919	31,110	199,029

Source: Ministry of Forests Annual Report 1996–97.

These figures tend to understate the degree to which even-aged management practices are used in BC, because of the way in which the Ministry of Forests defines these terms. In this tally, clearcutting includes only conventional clearcutting methods. Partial cutting includes patch cutting, coppice, seed tree, shelterwood, selection, clearcutting with reserves and commercial thinning, many of which are actually forms of clearcutting.

At present the dominant silvicultural system is still the even-aged clearcut system, but this may be changing. Opposition to its indiscriminate use is intense and growing. As well, under the new forest practices and land use laws, certain sites have been designated as requiring more sensitive management techniques than the even-aged clearcut system involves. And, finally, the composition of second-growth forests is such that selective harvesting as part of an uneven-aged management approach is fast becoming economically viable in some stands. In 1998, MacMillan Bloedel announced its intention to abandon clearcut logging and instead use a system called "variable retention" harvesting. This involves partial cuts of varying density depending on site conditions. In 1999 a few other coastal forest companies announced similar plans.

By the late 1990s the size of individual clearcuts had been reduced. According to convention, any forest opening of more than 1 hectare is classified as a clearcut. Regulations introduced in the 1990s limit the size of a clearcut to 40 hectares in the Vancouver, Kamloops and Nelson regions, and 60 hectares elsewhere, and the limits may be exceeded under special circumstances. Few foresters would argue that these restrictions, in themselves, engender a superior kind of forestry, but in lieu of any better, widely acceptable strategy, they have been adopted.

Some even-aged systems employ partial cuttings to remove a portion of the trees in one operation, and the remainder at a later date. The primary purpose of these methods, including shelterwood cuts, seed tree cuts and patch cutting, is usually to encourage natural regeneration on the site. The ultimate objective is still the creation of an even-aged stand, and these practices are really only a variation on, or a prelude to, clearcutting.

Some methods of partial cutting entail the removal of groups of trees in strips or patches, to create a stand of limited-age classes, or an uneven-aged stand. In practice most uneven-aged forests of any extent will, in time, present a wide range of silviculture options, including clearcuts. One would expect to find areas of even-aged trees, some of a single species. Other areas will contain various species, and/or areas with trees of many ages. Over time, the most dedicated proponent of selective harvesting will encounter situations where it is sensible to clearcut a part of the forest to control a disease such as root rot, to replace inferior growing stock, to provide wildlife habitat, to introduce a new or absent species, or any one of a hundred reasons. Proposals to ban clearcutting often fail to recognize these circumstances. A forest in which absolutely no clearcuts were allowed would be no healthier ecologically than what we have today, and it would be much less valuable economically. Foresters in other countries have even refined the terminology to reflect this fact. In Scandinavian countries, for example, permanent removal of a stand to convert the forest to another use is referred to as "clearcutting." But if a stand is totally removed for the purpose of reforesting the area, it is called a "regeneration harvest."

Many foresters unequivocally defend the use of clearcutting on both ecological and economic grounds. They maintain that clearcutting mimics natural disturbances, such as fires, in which forests are destroyed and regenerated. There are flaws in this argument: fires do not usually consume the same amount of timber removed in a logging operation; fires rarely kill all the trees in a stand, as clearcutting usually does; and fires do not disturb the soil, riparian zones and other forest attributes to the same extent or in the same ways as a conventional clearcut logging operation. On the other hand, the claim that all clearcut logging threatens biological diversity is also flawed. No one has yet been able to identify a single species that has been extinguished in BC by clearcut logging. It is possible to find areas of forest, clearcut in the past, in which certain plants have not regenerated and certain animal species have not reappeared. But it is equally possible to find second-growth stands as diverse and healthy as the old-growth forests removed from the site.

As well, matters of scale, timing and methodology all come into play. Clearcutting an entire watershed is quite a different matter than clearcutting a number of smaller areas within the watershed. Clearcutting a stand before adjacent stands have matured sufficiently to provide habitat for resident wildlife is different than implementing a harvest schedule that provides habitat for all species living in the landscape. Using grapple yarders, which require many roads, to clearcut steep slopes will have a different impact than clearcutting the same slope with a skyline logging system.[2]

Many foresters, and even more accountants and forest company executives, will argue that it is too expensive to use any system other than clearcutting. This is sometimes true. If the only consideration is the cost of logging, then clearcutting is almost always the most economical method. But if we also consider the cost of restoring the productivity of the forest and protecting or rehabilitating its non-timber aspects, then selective extraction may well be more economical.

For many people in the industry, particularly those who earned their stripes in the free-wheeling era of the 1950s and 1960s, arguments to persist in large-scale clearcutting are merely an excuse to carry on business as usual. The most vociferous proponents of the old way are seldom loggers, who have historically demonstrated adaptability and inventiveness. Given a clear set of parameters and the time to develop and test techniques, the average logger will not find the challenge of devising a cost-effective selective logging system insurmountable. If he thinks he will be able to stay in business long enough to recoup his investment, he will come up with an idea for the right piece of equipment, find an inventive iron bender in a local machine shop to build it, and eventually make it work efficiently and effectively. That is how most of the existing logging systems were developed, and how new ones will likely develop. Because clearcutting has for so long been the system used almost exclusively in BC, few loggers here currently possess the skills and equipment required to log selectively in an economically efficient manner. But as more selective harvesting is done and loggers develop expertise and equipment, these costs will drop.

Other factors also come into the equation. Ten or fifteen years ago, mills were reluctant to buy short logs. Today many mills prefer them. Bucking trees into shorter lengths reduces the cost of selective logging and makes it easier to avoid damaging trees left to grow. Bucking into shorter lengths can also bring in prices 35 or 40 percent higher for the same timber because of the higher grades achieved. This transfer of value from the mill to the forest can more than cover the added costs of harvesting a stand selectively.

Some BC forests have to be clearcut if they are going to be used at all for timber extraction. For instance, a climax coastal cedar–hemlock forest has evolved because of particular conditions of soil and climate. The two dominant species regenerate and grow well in the shade. Rainfall and other climatic conditions discourage fire, which would otherwise clear the site and allow the growth of species such as fir and pine, which need direct sunlight to become established. Typically these forests contain some very large cedar trees, a thousand years old or older, and a lesser number of large hemlock, which normally do not live so long. Between and beneath them grow a range of younger individuals of both species, perhaps with a few other shade-tolerant species mixed in. Removing any of these large trees, which are obvious candidates for harvesting by a selective cut, is an extraordinarily difficult task requiring the use of large, powerful logging machines. The damage inflicted on the residual stand is substantial, and it is impossible to avoid degrading the site. Such forests could be logged selectively, if somewhat expensively, but what would be gained? We would still be left with a cedar–hemlock climax forest, as it would be difficult or impossible to introduce shade-intolerant firs and pines into even a partially cut forest of this type. Thus, one of the primary objectives of using a selective harvesting system—to maintain or increase forest diversity—could not be achieved. If the stand or parts of it were clearcut, shade-intolerant species could be introduced. The point is not that selective logging is never unsuitable, but that it is not a panacea.

In the Interior, where the terrain is not so rugged and the trees are smaller, selective cutting is often more feasible economically, and more partial cutting is done there. The same is true for some coastal second-growth forests. Along the east coast of Vancouver Island, the Sunshine Coast and other coastal areas, there are extensive stands of second-growth timber of merchantable age and size, and access roads are already in place. Many of these stands grow near populated areas and are highly visible, or possess other valuable attributes such as water, recreational amenities or unique wildlife habitat. Clearcutting and replanting these sites—the standard Forest Service management plan—would undoubtedly be profitable in the short run. But taking into account longer-range economic and ecological factors, it may be advisable to begin harvesting some of these stands selectively and extending the time between clearcuts, or convert them to uneven-aged forests.

One of the consequences of our over-reliance on clearcutting has been the way it has influenced most of our post-harvesting silviculture activity. Silviculture in BC evolved in a clearcutting context and will tend to be perpetuated through repetitive cycles of growing and harvesting even-aged forests.

The purpose of silviculture, as defined by the Yale silviculturalist D.M. Smith, is "the creation and maintenance of the kind of forest that will best fulfil the objectives of the owner."[3] In BC, where the provincial government is the primary landowner, it is often difficult to discern the owner's objectives. At times the primary objective of silviculture expenditure appears to be the elevation of the allowable annual cut through the mechanism of the allowable cut effect; at other times it appears to be the relief of unemployment, or the wish to avoid offending political allies. The most consistent and explicable silvicultural activity has been the effort over the past two decades to restock logged-off and burned-over lands. Other than this, silviculture activity in BC seems to suffer from a lack of focus and direction.

Realistic silvicultural objectives must take into account the conditions of a particular forest as well as the desires of its owner. Under our centralized, state-owned system, basic decisions about silviculture investments are based not on a knowledge of the forest but on factors external to the forest, and they are arrived at through the application of standardized procedures devised by bureaucrats in Victoria who have never seen the forests involved. Consequently these decisions often are misguided.

Silviculture proceeds from the assumption that a managed forest is more productive than a forest left on its own. But silvicultural effort does not automatically improve productivity. The right action must be taken at the right time, in an economically, ecologically appropriate manner, or else the investment is not returned. In BC we have not devised a way of measuring the success of our silvicultural programs. It takes many years for the results to be known, and by then those responsible for the work have usually moved on. Very few of those people involved—from the tree planter who puts a seedling in the ground, to the forester who supervises the planter, to the bureaucrat who decided in the first place what species of tree would be planted—will ever see the results of their work or be accountable for it.

Forest management focusses on timber production. This does not necessarily mean other forest values are ignored. Good forest managers understand that the highest timber values will be obtained from the most productive forests, and the most productive forests are usually those in which a full range of plant and animal species associated with that forest site are sustained. But trees are the dominant and most obvious species in a forest, and trees are usually the most important economic product obtained from the forest, so most of what is known as forest management revolves around the growing of trees. The emergence of a more widely focussed type of forest management, sometimes called ecoforestry, holistic forestry or sustainable forest management, challenges the timber-biased approach and looks to the maintenance and enhancement of the entire forest ecosystem, including timber.

One of the most basic silvicultural activities is to influence the composition of a young forest. Decisions are made on what trees to cut, how to cut them, which species to encourage to grow and how to regenerate them. Individual malformed or defective trees must be taken out to expand the growing space of the more valuable individuals. Increasingly, forest managers are also concerned about the composition of non-timber species, some of which have a direct effect upon timber production. Brush that impedes the free growth of trees may be removed, while other brush species that provide winter forage for deer or elk might be retained and pruned to encourage more growth. Some foresters leave or establish snags to provide nesting habitat for woodpeckers, which eat bugs harmful to trees; other foresters erect birdhouses.

Forest management is beginning to include many other activities as well. One is the management of fish and wildlife habitat, although the fish and wildlife themselves are the responsibility of others. Water management is another aspect. The nature of the forest cover in a watershed has a crucial influence on water quality, quantity and rates of flow. Practices intended to enhance timber, wildlife, recreation or other values in a particular forest have to be managed so that they do enhance those values without compromising downstream water volume and quality. In some parts of the province, providing livestock range is part of forest management too. Far too often in BC, the relationship between silviculturalists and ranchers ends up being an adversarial one, though forests can be managed to enhance both timber and grazing values under the right circumstances.

Unfortunately most silvicultural opportunities have not been taken in BC. Until the 1920s there was no activity that could be called silviculture here, and before that time the only forest-related work not directly concerned with logging or protection was some preliminary experimental survey work conducted by the Forest Service. The first study of natural regeneration, an examination of Douglas fir seed distribution, took place in 1922. Three years later, regeneration studies were conducted on Douglas fir and western hemlock logging sites.[4]

Almost from the outset, there were two approaches to forest management—that undertaken by the Forest Service on 95 percent of the province's state-owned forests, and that practised by private owners of forest land. Most of the initial efforts started on the highly productive forest lands of Vancouver Island.

In 1924 the province's first reforestation project was launched when the government planted 7,500 deciduous trees near Oyster

River to serve as a firebreak. That same year, the Forest Service selected the site of its first experimental station at Aleza Lake in the Cariboo. In 1927 the first nursery, a temporary one, was set up at Saanich. Two years later a permanent nursery was established at Green Timbers, in Surrey. It began raising Douglas fir seedlings from selected seed sources.

By 1932 sufficient planting stock was available for the first production plantation, and a 221-hectare site of Douglas fir on West Thurlow Island was planted. In 1935 the Elk River Timber Company at Campbell River planted 16 hectares of logged-over land with seedlings it raised itself. That year 50,000 seedlings of Green Timbers nursery stock were planted. By 1938 the area planted each year had grown to 65 hectares.

That fall at Campbell River, 30,375 hectares of forest were destroyed by what was apparently the most expensive fire in BC's history to that point. In response, a new nursery was established nearby and production at Green Timbers was stepped up to provide enough seedlings to plant 4,500 hectares a year in the highly productive Sayward Forest. Prior to this, most of the planting had been done through a Depression-era unemployment relief program. With the advent of war and the sudden shortage of workers, tree planting was assigned to conscientious objectors.

The first major planting projects revealed one of the fundamental weaknesses of the BC system. Forest Service bureaucrats in Victoria decided the trees should be planted on a 6-foot grid, but many of the field staff disagreed, arguing for a 10-foot spacing. They cited the experience of silviculturalists in Washington and Oregon, who had established Douglas fir plantations two to three decades earlier. Their advice was ignored, and seedlings were planted 6 feet apart in most of the Sayward Forest. Thirty years later, with rapidly growing fir trees crowding the site, the Forest Service had to pay for an expensive spacing program to maintain growth rates. The first sad lessons of bureaucratically directed silviculture had been learned.

Meanwhile, during the latter half of the 1930s, some of the major forest companies with private holdings on Vancouver Island began to manage their forests for subsequent timber crops. The H.R. MacMillan Export Company, which had acquired its first timber lands in the Ash River valley near Port Alberni, adopted a patch logging system to encourage natural regeneration. Instead of logging from one end of the valley to the other, the company reduced clearcut sizes—to very large cutblocks, by today's standards—and left areas of forest to function as a seed source for the logged areas. Elsewhere on Vancouver Island, Bloedel, Stewart & Welch and a few other companies began replanting their logged-off lands. This was a marked departure from previous practice, which had been to cut timber and allow title to the land to revert to the Crown. The companies had made the decision to maintain permanent forests to supply their mills.

After the war, in his first Royal Commission report, Gordon Sloan was very critical of the 400,000 hectares of unsatisfactorily restocked forest land on the coast, as well as a substantial area in the Interior. He recommended it be reforested within twenty-five years. The Forest Service responded by opening more nurseries to provide the required seedlings. But it is far easier and cheaper to grow seedlings than to put them in the ground. When the government refused to provide funds for planting, the program had to be scaled down and some of the nursery plans cancelled.

At this time a nursery was opened at Elko, in the east Kootenays, to grow Douglas fir and yellow pine seedlings. To keep costs down, the Forest Service decided to use planting machines, which could operate only on flat lowlands. But ranchers had acquired grazing leases on these lands and their vigorous opposition to restocking them with trees brought the program to a halt. "Evergreen trees take too much out of the soil," one of the ranchers told Chief Forester C.D. Orchard. "They ruin grassland; they poison the soil. Evergreen trees and foresters are lice on the face of the earth."[5]

With the establishment of Forest Management Licences in 1948, the private sector assumed a larger responsibility for tree planting and by 1953 private companies were planting more each year than the Forest Service. Their motivation stemmed in part from the yield regulation system adopted in 1948 as part of the policy of sustained yield. Planting unstocked areas soon after logging, instead of waiting for them to regenerate naturally, triggered immediate increases in the allowable annual cut.

Annual plantings increased steadily, and in 1965 the Forest Service announced that in the future at least one-third of logged lands must be replanted—a policy requiring 75 million seedlings a year. Production of seedlings was rapidly expanded and by 1975, 91 million trees a year were being planted. By 1985 the annual planting was 100 million seedlings and in 1990 almost 300 million trees were put in the ground. The trend has been steadily away from natural regeneration to planting, particularly in recent years. In 1988, for instance, about 50 percent of all logged areas were planted; in 1993 that portion had increased to 65 percent.

This rapid increase in planting is not without its down side. Natural regeneration of forests has certain advantages. First, the high costs of planting—$465 a hectare in 1993–94—are avoided. Second, naturally regenerated forests usually contained more genetic diversity and a more representative selection of species. Early plantation forests tended toward monocultures because fewer species were planted than the original forest contained. Also, the seedlings used were drawn from a smaller seed source than natural ones on site.

These shortcomings have been reduced during the past twenty years by obtaining seeds from wider sources and by planting a mix of suitable species on a given site. A 1992 Ministry of Forests report said that plantation forests established on logged sites between 1970 and 1990 contained more tree species than the natural forests they had replaced.[6] As well, natural regeneration is less certain and usually slower than managed planting, which can lead to a lower AAC. The spacing of trees is less regular and the mix of species less predictable. From a timber production perspective, planting offers many advantages.

Until recently the Forest Service regarded silviculture as an option to exercise after an area was logged or burned. It had less control over the timing of logging operations and less opportunity than private landowners to treat logging as a part of a broader forest management regime as, for example, the MacMillan company did in the 1930s. Today most forest harvesting in BC is looked upon as one phase in a comprehensive silvicultural system.

Until the early 1970s, almost all silvicultural work on Crown land in the province was focussed on forest regeneration, and most of the energy and money were spent on planting. Beginning in the mid-1950s, some of the coastal forest companies operating Tree Farm Licences undertook the first stand-tending activities, including juvenile spacing, and brushing and weeding plantations. By 1977 they had spaced a total of 43,000 hectares and brushed 5,000 hectares.

The Ministry of Forests launched its first stand-tending program in the early 1960s, using prison inmates and unemployed workers. Although this work consisted mostly of spacing young stands, it was undertaken more to deal with aesthetic concerns than to enhance forest values. About 350 hectares of coastal forests a year were treated under this program up to 1977. In 1975, the first stand-tending projects were started in the Interior, with 170 hectares a year spaced in the east Kootenays. Aerial fertilizing was first tried in 1963, and two years later about 600 hectares a year were fertilized, almost all in coastal Douglas fir stands.

Beginning in 1978, these silvicultural activities were expanded substantially. An intense campaign by foresters and forest companies convinced the two senior levels of government that major spending programs on silviculture could significantly improve commercial timber yields. Over the next fifteen years, federal and provincial governments allocated several hundred million dollars to various intensive forest management programs. In addition, from 1978 to 1987, the costs of silvicultural treatments carried out by forest companies on Crown lands were rebated through credits against their stumpage payments. The major accomplishment of this period was the reforestation of a huge backlog of unsatisfactorily restocked lands denuded by logging and fires.

SILVICULTURAL TREATMENTS: AREA PER YEAR
(hectares)

Treatment	1960	1970	1980	1990	Cost/ha 93/94
Planting					
Area	4,000	12,000	63,000	200,000	$465
Seedlings	10 mill	25 mill	74 mill	245 mill	
Brush/weed	(1963–77: 6,500)		2,700	58,000	$317–$448
Spacing	(1963–77: 50,000)		22,400	22,000	$479
Fertilize*	(1963–77: 12,000)		5,500	3,400	$156
Pruning	Less than 2,000 ha to 1991, when 2,500 ha pruned, 3,000 in 92.				$880
Commercial thinning	(1963–77—nil)		30	150	N/A

*In 1986–87, 20,000 ha were fertilized, and until 1989 about 15,000 a year on average. In 1990, area fell to about 3,400 ha, increasing to 7,000 ha/year in 1991.

Source: Ministry of Forests.

In 1995 the second federal–provincial forest development program (FRDA) ended. Like the one before it, this program provided funds for the Ministry of Forests and forest companies to undertake approved silvicultural projects. Together they spent $500 million on silviculture treatments and related activities. By the conclusion of the second program, Forest Renewal BC was in place, with plans to spend about $26 million on silvicultural projects during 1995–96. Within a couple of years, the ministry's silvicultural budget had been off-loaded onto FRBC, which, in its first five years of operation, spent almost $450 million on silviculture. Critics were quick to point out that FRBC's costs per hectare were almost double those of the Ministry of Forests. By mid-1999, it was clear FRBC was a failure, and there was no indication where silvicultural funding would come from in the future.

Returns on silvicultural expenditures to date are unknown. In the late 1970s senior professional foresters claimed—with little or no supporting evidence—that they could obtain 300 percent increases in yield with intensive silviculture.[7] A US Forest Service study of silviculture in comparable Pacific Northwest forests concluded that with extreme care and attention, a combination of all available treatments might produce yield increases of 65 percent.[8] There has been little, if any, cost–benefit analysis of silvicultural expenditures in BC, but it is clear the extravagant promises of twenty years ago were never met. In the 1990s, in its explanation for reductions in AAC levels, the Ministry of Forests cited the failure of these silviculture programs to produce expected improvements in growth.

On the other hand, there is an enormous volume of timber available in mature and immature forests, most of which is not factored into AAC calculations and most of which is allowed to rot, which could be utilized by more intensive forest management practices. In second-growth forests this timber includes damaged trees, malformed trees, trees killed by insects or disease, wind-thrown trees, suppressed trees and the like. On the odd Woodlot Licence or privately owned forest, all or almost all the AAC is harvested from this wood. In other words, the annual harvest consists of salvaging natural mortality. This happens rarely, not because these conditions do not prevail in most forests, but because licensees have been unable or unwilling to establish the infrastructure of roads and landings to allow an annual culling of their forests. To some extent the experience of forest managers who have recovered this timber is a measure of the untapped supply available in the province's second-growth forests. That additional volume of timber, which could dramatically increase the annual harvest, is available for commercial use without impinging on current harvest levels and timber allocations.

The other large volume of timber available is the amount killed annually by insects and disease in old-growth or mature forests. Most of this timber mortality is caused by mountain pine beetles and western spruce budworms in the Interior, a yearly average of about 14 million cubic metres. Lesser amounts are lost to dwarf mistletoe infections and root disease, particularly on the coast. A considerable (if unknown) portion of this wood is salvaged during logging, but most of it is lost because it is found in isolated areas of old-growth forests into which roads have not yet been built.

In some cases these losses are the result of past forest policies and practice. In the Cariboo, for example, the primary pest is the mountain pine beetle, which infects older—some would say "overmature"—lodgepole pine forests. In the past, huge fires swept through these areas and the forests rarely reached the advanced ages at which they are most susceptible to beetle attacks. With our recently acquired ability to combat forest fires, these forests have become older and more vulnerable to insects. When an infestation occurs it can spread rapidly, consuming vast areas. In the 1970s a bark beetle infestation escaped from Bowron Lakes Provincial Park and destroyed most of the trees in a 20,000-hectare tract of commercial forest land to the north. A massive clearcutting operation ensued to salvage the timber and stop the infestation. The area was replanted immediately. Many people objected to the size of this clearcut, as they had earlier to suggestions that the infected, dead trees in the park be salvaged before they touched off the larger infestation.

There are dozens of potential Bowron situations throughout the Interior; under the right conditions they could explode. A similar, smaller infestation occurred in the west Chilcotin a few years after the Bowron loss, and most of the forests in that region are now infected with beetles to some degree.

The option of harvesting infected trees before bug attacks reach epidemic proportions has not been available until recently. Roads are normally built as timber harvesting proceeds into unlogged areas, so it has been impossible to reach infected pockets of trees in remote, undeveloped areas. More recently, the ability to detect these attacks at early stages has improved, and helicopter logging is economically feasible. It is within the realm of possibility, given a period of good markets, to surgically remove these infected pockets to prevent the spread of disease, and to salvage useable timber at the same time. Here again, a more intensive kind of management could improve forest health and provide additional volumes of timber to pay for it and expand economic opportunities.

The major impediment to the development of a forest management capability in BC that goes beyond basic reforestation is the overall policy framework and structure of the forest sector. Since the early 1970s BC has attempted to acquire the ability to enhance the timber supply efficiently and effectively by applying well-known silvicultural techniques developed elsewhere. In this, BC has failed. More recently people working with forests have come to understand the need to manage them for the maintenance and enhancement of non-timber values, to become competent stewards of entire forest ecosystems. Given existing policies and structures, this new understanding cannot be applied.

The policies that now determine the structure of the industry were designed to facilitate the development of a timber harvesting and processing industry. With difficulty, a reforestation capability was developed later by making it a requirement of retaining the right to harvest Crown timber. Still later, in an attempt to extend silvicultural efforts, vast sums of (mostly public) money were spent on attempts to enhance the timber supply. In the end the undertaking proved inefficient, and governments turned out to be unreliable sources of funding for this type and scale of forest management.

Because the prevailing policies provide no incentives for licensees to engage in fully developed ecosystem management, the NDP government of the 1990s attempted to legislate this capability into existence through a command-and-control system of regulating forest practices. This approach has turned out to be extremely expensive and, so far, apparently not very effective in maintaining or enhancing the full range of values found in BC forests.

Until the 1990s, most silviculture undertaken to maintain and enhance the province's forests was done under short-term contracts administered by the Ministry of Forests and the major forest companies—by some 18,500 seasonal workers employed for an average of three months each. Most of these workers were temporary, untrained, transient employees of small silvicultural contracting firms. Among them was a core of experienced workers who had acquired the skills upon which the future of the province's forest economy rests. Members of this core group worked anywhere from three to ten months a year, chose not to belong to a union, and had little job security.

In the mid-1990s, when responsibility for silvicultural funding shifted from the Ministry of Forests to Forest Renewal BC, the silvicultural workforce suffered a major disruption. Previously a well-established bidding system awarded silvicultural contracts to firms employing experienced workers who were not union members. Preference in FRBC funding was given to silvicultural operations employing unionized loggers and millworkers who had lost their jobs as a result of various government forest sector programs that reduced the timber supply. Experienced silvicultural workers were unable to find work, and in many cases did not qualify for FRBC-funded jobs as displaced forest workers. In 1997 the government announced all silvicultural workers employed under FRBC funding must belong to a union.

That year, an FRBC subsidiary, New Forest Opportunities Ltd. (NewFo) was established. Its mandate was to function as a hiring hall, employing all silvicultural workers and assigning them to silvicultural contracting firms, which would not be permitted to hire non-NewFo workers. All NewFo workers had to be or become members of IWA Canada. In its first year of operations, the NewFo plan applied only to coastal silviculture projects, including spacing, pruning, watershed restoration and recreation management. The plan was to extend the program into the Interior and to reforestation workers throughout BC at some date in the future. A year later, after checking the results of the first year's operation, the forest minister admitted the NewFo scheme was too expensive to expand as intended.

In practice, most silvicultural training is undertaken by contractors, and on-the-job costs are borne by them and by workers motivated to learn the necessary skills. In 1993, the BC government established the Forest Worker Development Program to train income assistance recipients and others in a variety of forest-related jobs, including silviculture. The objective was to train workers, who could then find work with silvicultural contracting companies. In its first year the program cost $34.5 million and provided about 3,200 people with twelve days of basic training each. The $900-per-person-per-day cost also covered a limited amount of actual silvicultural, recreational and range management work.[9] Silvicultural contractors estimate that after three years of the program's operation, only about 5 percent of the trainees ended up steadily employed in the silviculture sector. The real beneficiaries of this government largesse were the small army of bureaucrats and consultants who administered the program and provided training services.

Part of the reason for the instability of the silvicultural sector is the way silvicultural work has been financed. Under most tenure arrangements, except for those conducted through the Small Business Forest Enterprise Program, it was the responsibility of whoever logged an area of Crown land to reforest and bring it to a free-to-grow state. Beyond that, most silvicultural activity—essentially stand-tending practices aimed at enhancing timber growth and quality—was paid for by the licensee. The Forest Service paid for SBFEP reforestation and for some additional silvicultural activity on Crown land.

SILVICULTURAL WORK, 1996–97
(hectares on Crown land)

Funded by:	BC MoF (SBFEP)	FRBC	Forest Companies
Site preparation	11,112	743	79,574
Planting	24,193	1,547	149,924
Brushing	3,573	4,664	35,312
Spacing	94	32,248	4,141
Fertilizing	74	14,804	4,559
Pruning	910	7,673	32

Source: Ministry of Forests Annual Report, 1996–97

As the numbers indicate, in 1996–97 most private sector silviculture on Crown lands was the sort required to maintain the right to harvest timber, with little done to enhance future forest crops. Evidently forest companies undertake their contractual obligations responsibly and efficiently, but they are not confident enough about receiving the benefits of other silvicultural activity to spend significant amounts of their own money on it.

Through the late 1980s and early 1990s, the Ministry of Forests silviculture budget was more than $200 million a year, much of it from the joint federal–provincial FRDA program. The Socred government off-loaded some silvicultural costs onto industry in 1987 and, beginning in 1995 when FRDA ended, the NDP government began off-loading costs onto Forest Renewal BC and shut down its silviculture branch. By 1997 the ministry was spending only about $60 million a year on silviculture.

As conceived in 1994, Forest Renewal BC collected money in the form of a Super Stumpage assessment when lumber prices were high and processed it through its own internal bureaucracy, which ate up about $20 million in 1997. Then FRBC contracted with either the government or the private forest companies, who in turn hired contractors or organized crews made up of workers laid off by mills and logging operations. Beginning in 1998, these workers were employed by NewFo.

At first FRBC silvicultural funding was to supplement Ministry of Forests silvicultural spending, but as the ministry's silviculture budget plunged, FRBC expenditures came nowhere near replacing that budget, let alone paying for additional silviculture. When FRBC was established, the government announced the new agency would spend $100 million a year on forest management enhancement programs. FRBC announced in early 1997 that it intended to spend $56 million during 1996–97 on silviculture, which included many of the training functions previously performed under the costly Forest Worker Development Program.[10] Combined with the ministry's reduced silviculture budget, silvicultural expenditures that year were in the $100 million range—less than half of what they had been a decade earlier when the work was being done by a relatively experienced contracting sector. By the end of 1997 the silviculture sector appeared to be headed into a state of semi-chaotic decline, even though FRBC spending increased substantially. Within a year or so, the consequences of reorganizing the silvicultural apparatus were apparent. Independent but reliable estimates concluded that on a per-hectare basis, silvicultural costs had almost doubled and real accomplishments declined. It was also suspected that the quality of work had degenerated.

In 1999, FRBC's new business plan indicated that the agency would go through its financial reserves within two years, and if forest product prices were not high enough to bring in additional revenues, FRBC would be out of money. This and other indicators suggested the agency would not survive past a couple of years, and there was no indication of how silviculture would be funded and administered beyond this point.

On private forest lands, the silvicultural spending process works quite differently. It is treated as a straightforward investment in a future timber supply. Money earned in private forests does not have to be cycled through large bureaucracies before a reduced portion of it finds its way back into the woods. As well, forest companies invest more on their private lands than they do on Crown lands they lease. A study of the amounts invested in silviculture by the private sector under various forms of tenure indicated that for every $1 spent on silviculture on a Forest Licence, $1.14 was invested on Timber Licences, $1.24 on Tree Farm Licences and $1.73 on private land.[11] Clearly, willingness to invest in silviculture grows as security of tenure increases.

Funding of silviculture is an erratic affair, undertaken grudgingly by both government and industry. Under these circumstances, forest companies are not willing to hire permanent silvicultural workers; nor is the government. The only permanent members of the silvicultural workforce in the province are professional foresters and technicians in government and industry whose job is to plan and supervise the temporary workers who do the actual work in the woods. Except for the relatively small number of woodlot licensees and small private forest land owners, there are no opportunities for individual forest workers to make silvicultural investments of their own. As a result, apart from a small core group of contractual silvicultural workers whose security and organization has been undermined by FRBC and NewFo, there has been no accumulation of silvicultural capability in BC. We have become experts in the erection of silvicultural bureaucracies but are still at a primitive level of silvicultural expertise.

On virtually all government-owned land and private land included in government-regulated tenures, silviculture is practised in a series of separate phases or treatments that are performed in a fairly uniform manner over relatively large areas as defined by individual contracts. Under the Forest Practices Code, the detailed execution of these treatments is prescribed in a series of guidebooks. One of the major activities and expenses associated with silviculture in BC occurs not in the woods but in the office. A convoluted planning process is required and various approvals and permits must be obtained before any work begins. Growth of the silvicultural bureaucracy over the past two decades has exceeded increases in actual silviculture many times over.

Under law and Ministry of Forests regulations, silvicultural work starts before an area is logged. A Silvicultural Prescription (SP) must first be prepared for each cutblock. Under current regulations, cutblocks normally cannot exceed 40 hectares on the coast or 60 in the Interior, and most are much smaller. With more than 200,000 hectares logged annually, as many as 10,000 SPs must be prepared and approved each year.

According to the 72-page guidebook detailing the form and contents of a SP: "The Silviculture Prescription prescribes the method for harvesting the existing forest stand, and a series of silviculture treatments that will be carried out to establish a free growing crop of trees in a manner that accommodates other resource values as identified. Subsequent documents, including cutting authorities and logging plans, must follow the intent and meet the standards stated in the silviculture prescription."[12]

This is a legally binding document, requiring the signatures of the licensee, a Registered Professional Forester and the district manager of the ministry before logging can proceed. The first step is to prepare a detailed administrative and physical description of the site. It must describe all the timber and non-timber attributes of the site and explain how they will be affected and, in some cases, protected. These attributes include fish, wildlife, water, recreational opportunities, visual landscape values and cultural heritage resources.

The second step is to describe the silvicultural system to be used, beginning with an explanation of how it will achieve forest management objectives for the stand as well as for the wider area in which it is located, including meeting objectives for biological diversity, visual quality and timber production. The applicant must define the harvesting system to be used, along with any constraints on logging, and means of reducing fire hazards must be included. The ministry provides a separate guidebook for the selection of silvicultural systems.

The third step is to describe how the area will be restocked after logging and how it will be managed after planting until the seedlings have reached a stage where they can grow freely on their own. This includes methods of keeping the site clear of competing brush, of spacing the newly established stand, of pruning the new trees, and of fertilizing if necessary.

As in the rest of the Forest Practices Code, the emphasis in the design and approval of SPs is on the process, not the final result—i.e. the state of the forest that replaces the one logged. It is not as if a forester employed by the licensee and another forester employed by the ministry go to the area proposed for logging and, after careful deliberation based on their knowledge, training, experience and familiarity with the site, arrive at a decision on the objectives of the operation. Instead the SP is a document carefully crafted to answer all of the questions raised in the Ministry of Forests guidebook. Under the law, that is all that is required. If the terms of the SP are adhered to and the new forest turns out to be a silvicultural disaster, no one can be blamed. In other words, when good silviculture occurs as a result of this approach, it will be in spite of the regulations and not because of them. The process is rigorous but at no point does it say the bottom line is to leave a healthy forest on site, or that the end result is what the licensee will be judged on. It just says, "Jump through all these hoops and you're off the hook."

After the stand has reached a free-growing state, another document, a Stand Management Prescription (SMP), is drawn up. It is the authorized plan for all silvicultural activities from the achievement of a free-growing state to the drawing up of the next SP, which could be as much as a century in the future. Although licensees are not obligated to implement the SMP, they are prohibited from undertaking any silvicultural activity not defined in the SMP. Amending an SMP requires the signature of the Ministry of Forests district manager.

The SMP defines management objectives for the various timber and non-timber attributes of the stand. It includes silvicultural, forest health, fire protection, soil conservation, fisheries, wildlife and other objectives. More detailed procedures for meeting these various objectives are contained in a long series of ministry guidebooks and the information in the SMP must be recorded on a number of government forms.

Like the SP, the SMP is focussed on a regulatory process, rather than the economic and ecological well-being of the site. Both of these provisions of the Forest Practices Code can end up forcing licensees to engage in silvicultural activity they were unwilling to invest in on their own initiative because of the lack of secure tenure, and they can lead to a well-earned distrust of government. And there are many more such regulations. The effect through 1996 and 1997, when these regulations came into full effect, was to raise logging costs substantially, and to cause logging operations to be postponed or cancelled because of a backlog of paperwork. By mid-1997, under extreme pressure from industry and unemployed forest workers, the government frantically attempted to reduce the regulatory burden, while environmentalists worked harder than ever to resist what they perceived as a return to poor forest practices.

In the long run, there has been some real improvement in silvicultural practice since the 1970s. Until the 1990s, silvicultural activity focussed on a narrow range of options that consisted essentially of monocultural forestry. At best, this approach simply ignored forest diversity; at worst, it deliberately sought to destroy it. Most reforestation involved selecting the most commercially valuable species that might grow on a site and planting that species in vast monocultural plantations. By the late 1990s, mixed species were being planted on most sites and there was a conscious attempt to maintain or enhance the original diversity of tree species on the site. As recently as a decade ago, spacing activity worked toward the same monocultural objectives. The less desired species—usually the least commercially valuable—were systematically removed. Now, spacing workers are expected to take an ecological approach to their job and consider the diversity of all species on site.

The main problem is that for so many decades silvicultural work focussed obsessively on enhancing timber volume, rather than timber value, not to mention non-timber forest values. Spacing and fertilizing were attended to with vigour, while virtually no money was spent on such value-adding work as pruning. The situation had improved slightly by the end of the 1990s, but the figures indicate that the emphasis is still on volume enhancement.

Still, we have barely begun. If we are going to have a healthy forest-based economic sector in this province in the future, it must consist of an industry that utilizes timber produced or improved by human effort. The scale and vitality of that industry will be in direct proportion to the quantity of timber we grow, as well as its quality. The Ministry of Forests has stated repeatedly that with our current level of silvicultural activity the volume of timber available to industry from Crown land will decline to a little more than 50 million cubic metres a year. With a fully developed silvicultural capability, that volume could be doubled or even tripled in time, and its quality could be improved as well.

To compete in the global forest products market, where we now sell 80 percent of our production, British Columbians must become very good at the business of growing trees. To continue operating, without being restricted unduly by the people of BC and our customers around the world, we must learn to manage forests for all of their values, not just timber. To succeed at this, to become true stewards of the entire forest, we need to acquire a silvicultural capability far beyond what we have now.

This will not happen simply as a matter of course through some natural evolution of forest policy and practice. A major revision of BC's forest policies and a significant restructuring of the forest industry must take place. These changes must take place within the context of defining and agreeing on some broader social objectives. If British Columbians agree to strive for a healthy, expanded forest economy based on an ecologically sound concept of forest stewardship, which supports a socially and economically diverse industrial structure, the task of reorganizing the industry and learning how to become good forest managers will be much simpler than it has been until now.

The Use of the Forest 10

A new concept of forest administration appeared in British Columbia in the early 1940s, just before the appointment of the first Sloan Commission. The concept of multiple use—the idea that a forest could have more than one user—had been around since the beginning of the forest industry. But until Chief Justice Sloan was instructed to consider "the use and management of forest and wild lands for parks, recreation, grazing and wild life,"[1] the notion was an abstract one rather than a priority in practice.

When Sloan initiated his inquiry, the only forest use recognized in legislation was the production of industrial timber. In 1925, the Minister of Lands introduced legislation establishing Provincial Forest Reserves, stating that "the primary object of their creation is to ensure a continuous production of timber."[2] These provincial forests were legally defined areas placed under the sole jurisdiction of the Forest Service. At the time this was seen as good conservationist policy, an enlightened alternative to converting forest land to other uses such as farms, rangeland or wasteland after forests were logged.

As time went on, the Forest Service pushed aggressively for creation of Provincial Forest Reserves throughout the province, on all forested areas outside parks. Other government departments—agriculture and fisheries, for example—objected from time to time, but were easily overruled: timber production fattened the provincial treasury.

In his two Royal Commission reports, Sloan did little more than acknowledge the existence of other forest users and values. In paying scant attention to the needs and concerns of non-timber interests, he was merely reflecting a widespread attitude of the day. There had always been a few critics of industrial forest practices and their impact on other forest resources, chiefly fish and wildlife, but they had little effect on the industry or on forest policy.

Before 1956, the major resource use conflicts were with the commercial fishing and ranching industries, and occasionally with sport fishermen. A potential conflict with trappers and hunters never materialized: most trappers simply moved farther up watersheds. In the early days of logging, sport hunters were barred from the woods on the grounds they were a fire hazard, but after World War II the relationship changed. Many young second-growth forests support large deer populations, and logging roads provide hunters with easy access. Hunters' organizations became strong supporters of the forest industry.

The conflict between forestry and ranching had different origins. The cattle industry operates primarily in the ponderosa pine–Douglas fir forests of the southern Interior. Logging this country created rangeland so ranchers were not opposed to the timber industry per se. Reforestation and forest management, however, were another matter, as Chief Forester C.D. Orchard discovered when he attempted to plant logged-off lands in the Kootenays.

The BC ranching industry operates from a base of about 1.2 million hectares of privately owned grasslands, and an additional 11 million hectares of forested rangeland leased from the government. About three-quarters of a million head of cattle graze this rangeland annually. Most of it is covered by both grazing and forestry leases held by different licensees, and the potential for conflict is substantial. Cattle trample young seedlings, for instance, making it difficult for forest managers to establish a new forest. Foresters view unplanted land with the same abhorrence with which nature views a vacuum, an attitude ranchers see as a threat to their livelihood. No form of lease is available that permits the integrated management of both resources by a single licensee.

After World War II, with the rapid expansion of logging into remote areas, resource use conflicts grew too frequent and intense to be resolved in informal ways. One of the first major conflicts was with the salmon fishing industry, as logging practices began to damage salmon spawning and rearing streams. In 1956 the provincial Forest Service and the federal fisheries department instituted a "referral" system whereby the Forest Service agreed to inform fisheries officers of logging operations planned near important salmon streams and to consider incorporating those officers' suggested restrictions into logging practices. In 1970 the same referral system was extended to permit provincial wildlife officials to comment on logging proposals.[3]

In neither case did the referral system work particularly well. In the first place, ultimate authority to determine how an area would be logged lay with the Forest Service, which meant fisheries and wildlife concerns could be ignored. Also, with 2,000 or more cutting permits issued each year, the understaffed fisheries and wildlife agencies were unable to keep up with the work. Inevitably they made hasty and ill-conceived judgements, or they were unable to examine the site, leaving the Forest Service to proceed without a reply, assuming silence was consent. The situation was aggravated by a lack of reliable inventory data for affected sites.

As progressive clearcutting intensified and roads were pushed into rougher terrain in the early 1970s, conflicts with other resource interests, particularly fisheries, escalated. In 1972 the Forest Service issued a set of guidelines for coastal logging operations. Minimum standards were set in riparian zones to limit the size of cutblocks, protect streams, regulate road construction methods and so on. Soon after, similar guidelines were issued in parts of the Interior. The guidelines were an attempt to impose a set of standards on logging operations and to alleviate shortcomings of the referral system, which was still in force. But like any broad set of rules, the guidelines set out generalized requirements that often did not fit the conditions of particular logging operations. One of the basic weaknesses of this and other early attempts was a lack of competent professionals to diagnose, design and make good decisions. The guidelines either failed to protect non-timber resources adequately, or needlessly restricted loggers. For example, one standard directive was to leave a 30-foot strip of unlogged land along a fish-bearing stream, because debris and eroded soil can harm fish, and because tree cover is essential to maintaining water tempratures. But in some cases, 30 feet was insufficient to prevent stream siltation. In other cases, depending on wind patterns, trees left standing might blow down into the stream. The Forest Service itself had problems with the system, and in 1975 told Peter Pearse during his Royal Commission hearings that a site-specific, interdisciplinary planning system was needed.[4]

In 1973, during the reign of the Dave Barrett NDP government, the first planning process of this type was introduced. It was known as "resource folio planning." The system took its name from the portfolio of maps used to illustrate the available soil, forest cover, wildlife, fish, recreation and other kinds of data on each resource for a given watershed. These maps were printed on transparent material and could be overlaid to indicate actual and potential interactions or conflicts. The Forest Service sought comments from other government agencies on how each area should be managed and then worked with logging companies to design a five-year cutting plan for the area.

The folio system was developed on the assumption that different forest values and user groups could be accommodated through a detailed advance planning system. These values had been looked after previously by general guidelines—a series of thou-shalts and thou-shalt-nots intended to regulate the behaviour of a particular user, usually the timber industry. Under the folio system, the emphasis shifted from the operational level to the planning level and, naturally, led to a large increase in the number of planners employed in various resource bureaucracies.

The evolution of multiple-use management was given great impetus by Pearse's recommendation that forests should be managed for more than their timber resources. His advice was incorporated into the 1979 Ministry of Forests Act, which explicitly instructed the Forest

Service to consider all forest resource values. The new goal was to maximize the benefits obtained from forest land by combining or integrating management of several resources, instead of maximizing a single resource. On multiple-use lands, constraints would be placed on activities such as timber harvesting that compromised other resources.

In practice, multiple-use management had severe limitations, at least in its early stages. This was partly because of certain definitions. Single-use allocation of forest land, unless it bestowed overwhelming benefits to society, was eschewed in favour of multiple use. However, logging and forest management were, by definition, seen as one of many potential uses for most forest land. Extracting timber and restocking the forest with trees was not seen as incompatible with most other uses. Wilderness preservation was seen as a single use, comparable to flooding a valley by building a hydroelectric dam, because it precluded many other uses such as logging and mining. By the late 1970s, provincial forests had been defined as multiple-use land, while other land such as parks and wilderness, agricultural or settlement areas were considered single-use land.

The timber bias of multiple-use planning was exacerbated by the Forest Service's domination of the system. Multiple-use planning was not a process in which all public resource agencies sat down together as equals, with or without private sector forest users, and decided how an area of forest would be managed. The old legislation that had established the Provincial Forest Reserves, as well as the Ministry of Forests Act of 1979, were jealously guarded by the Forest Service. By habit and inclination the Forest Service supported the industry's primacy in the land use planning process and viewed all others, including other government agencies, as interlopers.

In earlier periods the Forest Service had taken on other functions pertaining to certain non-timber resource uses in the provincial forests. Large areas of forest land, much of it within Provincial Forest Reserves, had long been used as grazing land by ranchers, and separate bureaucracies in the Lands Branch and Forest Service had evolved to administer it. Logic might dictate that the province's range management activities would best be handled by the Ministry of

Agriculture, but under the terms of the legislation establishing Provincial Forest Reserves, rangelands would then have to be moved out of those reserves. To prevent such a move, the Forest Service developed a Range Branch and, as recommended by Peter Pearse, took over range-related functions performed by the Lands Branch prior to 1979. Similarly, as recreational use of forests expanded, the Forest Service established a Recreation Branch, which developed hiking trails, campgrounds and other amenities on Crown forest land outside Tree Farm Licences.

In 1979 the Forest Service circulated a discussion paper within the provincial bureaucracies proposing that all unclassified forest land either be placed in a Provincial Forest Reserve, or be designated as single-use land and left out. This paper precipitated a huge jurisdictional squabble among bureaucrats, who will work as hard as any creature in existence to protect their turf. The dispute was eventually resolved through a series of agreements and protocols among government ministries, which defined administrativc rcsponsibilities and established procedures for removing Crown land from Provincial Forest Reserves.

The concept of multiple use entailed something quite different for Tree Farm Licences. From the time they were introduced in 1948 until 1979, there was no mention in the legislation of any other resource values except timber. Nor were any provisions for other resources included in TFL contracts. In 1979 an amendment to the act required licensees "to consider" recreational, fish and wildlife values if inventory information was available.[5] In practice this meant that if it was known that a small stream contained fish, and if that information was included on the appropriate official map, then some consideration had to be given to the welfare of the fish during logging. If it was not known whether fish used the stream, then no consideration was required.

In 1987 another amendment required TFL licensees to include "information on the inventory of the forest and recreation resources and, where available to its holder, of the fisheries and wildlife resources of the TFL area."[6] Now a licensee was required to take an inventory of an area's recreational capabilities, although there was no requirement to protect it. And there was still no consideration of fish, wildlife or other values. The amendment created more problems than it solved. A great debate between licensees and the Forest Service ensued, about which types of recreational data should be acquired, standards of data collection and, most important, who would pay for undertaking the inventory.

Any given area of forest land, for example a watershed, contains resources like timber, fish, wildlife, minerals, edible plants, water and so forth. It also has numerous potential resources such as unforested land on which timber can be grown, habitat capable of supporting fish and wildlife species that are not present, opportunities for recreational activities such as camping, hiking and canoeing, and other industrial activities such as grazing, mining and hydroelectric power. The concept of multiple-use management suggested that at least some of these resources or resource opportunities should be taken into account if any one resource user, such as a forest company, wanted to make use of the area.

However, the rules and regulations under which a forest company was permitted to use the area, whether for logging or long-term forest management, limited it to a single use—the production of commercial timber. The forest company had to incur costs if it accommodated other resources on the site, but had no means of realizing the benefits of caring for that resource, or any other incentive to do so. The usual forest company response was to ensure continued access to timber by meeting the minimal requirements of protecting non-timber resources. The system did a poor job of assigning responsibilities, costs and benefits.

Attempts to deal with this problem led to the emergence of another administrative concept, integrated resource management (IRM), through the late 1970s and early 1980s. This new concept sent the old idea of multiple use to the administrative trash bin. The principle of multiple use suggested that every hectare of forest land could simultaneously serve industrial forestry, water, wildlife, recreation and the like, but IRM was conceived as a process through which combinations of resource uses in an area could be identified and planned. It was meant to be flexible, opening the way for a number of new institutional arrangements such as interdisciplinary task forces and study groups set up to consider the use of particular areas of forest land.

One of the first groups of this sort was the Tsitika Planning Committee, established by the Environment and Land Use Committee (ELUC) of cabinet in 1977. The Tsitika River watershed encompasses about 40,000 hectares of heavily forested land on northeastern Vancouver Island, about 60 percent of it covered with commercially useful timber. It was the last major unlogged watershed in the northeastern part of the island. It has a rich mixture of wildlife, including elk, deer, wolves, cougar, bear and smaller fur bearers. The river contains one of the most diverse fisheries resources on Vancouver Island, with substantial runs of salmon and steelhead. The river estuary at Robson Bight is an important site for killer whales. The watershed and estuary includes two lakes considered to have high recreational values.

Almost the entire watershed had been allocated as portions of three Tree Farm Licences in the late 1950s and early 1960s, and by the 1970s the companies holding these licences were ready to start logging in the valley. They intended to build a road from the estuary into the upper reaches of the watershed and haul logs to a dry-land sorting yard they proposed to build at Robson Bight. The original TFL agreements, recognizing the non-timber values of the watershed, contained clauses allowing for portions of the area to be withdrawn for park and other non-timber uses, and in 1972 a loose coalition of non-timber resource agencies and public interest groups came together to oppose the logging plan. Lacking any other instrument of change, they proposed the entire Tsitika drainage be withdrawn from industrial use and placed into an ecological reserve, with parks established at the two lakes. These proposals went far beyond the intent of the withdrawal clauses in the TFL agreements. In 1973 Bob Williams, the NDP forest minister, declared a moratorium on logging in the watershed to allow time to consider the various use proposals.

The Forest Service established a study group to devise a set of land use options for the watershed, which were presented at public hearings. ELUC examined the options, along with the public's reaction, and in 1977 decided an integrated management plan should be prepared for the watershed. The task was assigned to the Tsitika Planning Committee, which initially consisted of bureaucrats from the provincial fisheries, forestry and wildlife agencies; the federal fisheries service; and representatives of the three TFL licensees. Later a local conservationist and representatives of the forestry and fisheries unions were added. They held intense meetings for a year and produced the Tsitika Watershed Integrated Resource Plan in October 1978. It recommended several measures to protect fisheries, wildlife and recreational values, all of which would reduce logging activity in the area.

Two points about the Tsitika planning exercise are worth noting. One is that once the process began, there was no end to it. The planning committee's report and recommendations hopelessly confused objectives for the valley with the means of attaining them, and were only the first step in a long process that was still in train almost twenty years later. In 1995, with the announcement of the Vancouver Island Land Use Plan, which included removing 3,600 hectares of forest land from the industrial forest base in the lower Tsitika, there was still no resolution on what to do in the watershed.

The second point is that in spite of the name of the process, it had little to do with the actual management of resources in the Tsitika, apart from forestry. In the Tsitika and elsewhere, IRM does not really refer to managing resources, but to talking about or planning resource management. More specifically, it deals with an aspect of forest management—timber harvesting—and its impact on other resources, and restrictive measures to protect those resources. IRM is a concept not so much of management as of administration, dealing with the integration of bureaucracies rather than the integration of resources. Since the introduction of the IRM concept in the late 1970s, bureaucratic processes have grown and very little actual work has been done at ground level. For the most part, IRM is little more than a sophisticated set of restrictions on the primary forest resource user, the timber industry.

Throughout the 1980s, despite the widespread dissemination of the idea of IRM, conflicts over forest land use intensified. As the rate of logging accelerated and the industry was forced into more remote areas to find enough timber to fill its allowable annual cut allocations, the number of disputes grew. The growing popularity of outdoor recreational activities, combined with an expanding environmental movement, brought even more pressure to bear on resource users, particularly the timber industry.

The Forest Service was caught in the middle of this conflict. On one hand it was attempting to fulfill the mandate it had been handed in 1979 to consider non-timber resource values. At the same time it had a deeply ingrained seventy-five-year habit of working as an adjunct to the forest industry. It was thrust into the position of putting the principles of IRM into practice without reducing AAC levels to the point where logging activity had to be curtailed.

In 1987 the Forest Service set up the Integrated Resource Management Program, combining its old range and recreation management programs with a new integrated resource planning function. The following year, after extensive study and consultation, a revised set of Coastal Fisheries/Forestry Guidelines was adopted. It was a much more detailed version of the guidelines developed in the early 1970s.

Another pressure on the Forest Service stemmed from the tendency of many of its senior members to dogmatically defend the status quo, reinforcing a growing public perception that the government agency was biased toward the forest industry. By 1990 the Forest Service was visibly divided over its priorities in the woods. That year BC forest minister Claude Richmond told a meeting of senior forestry managers in Victoria: "There are many values in our forests besides timber . . . The question in the forefront of public concern a few years ago was jobs, now it is the environment . . . I say to people in the industry that if you don't learn the new game and new rules you will not be playing very much longer . . . There are a lot of players in the industry who will not be with us because they will not change. They will not adapt."[7]

That spring the Forest Service's annual district managers' meeting focussed on IRM. In a follow-up memo, Deputy Minister of Forests Phil Halkett reinforced the message: "The Executive of the Forest Service fully supports IRM and expects field staff to demonstrate IRM philosophy while carrying out their duties . . . We need to demonstrate, at every opportunity, our commitment to provide equal consideration to all resource values, and restore public confidence in the Ministry [of Forests]. The public should perceive the Ministry as 'value neutral.'"

Wes Cheston, the assistant deputy minister and head of the operations division of the Forest Service, had also signed Halkett's memo, but he appeared to have missed the point. Cheston oversaw almost all of the administrative functions performed by the ministry in the forests. Six weeks after the memo was sent out, he appeared before the Forest Resources Commission and said, during an exchange with FRC chairman Sandy Peel: "Within a Provincial Forest the primary use of all areas is the growing and cropping of trees and those other uses [fish, wildlife, recreation and water] are secondary to that."[8] Peel was so astonished by this statement he asked Cheston to repeat it, which Cheston did, in even more forceful terms. Later that day, apparently after conferring with his superiors, he returned and retracted his statement, admitting that under the Forest Act a provincial forest can be managed for various purposes, including timber production.

Later that same year, the Forest Service issued a major policy paper on integrated resource management, which it defined as:

> a process which identifies and considers all resource values along with social, economic, and environmental needs. This process assigns resource use and management emphasis based on present uses, the mix of benefits, and social preference. IRM produces a mosaic of single, concurrent or sequential uses which may be few or many, and which meet agreed upon management objectives without unnecessarily impairing the productivity of the land. The approved pattern or sequence of use[s] provides the desired mix of benefits and may not necessarily provide the greatest financial return or the greatest unit output from any individual use.[9]

By the time the Forest Resources Commission released its report in April 1991, integrated resource management had acquired the status of a law of nature. For the commission IRM was a given, a fact of life, and it was barely mentioned. The FRC was much more concerned with the dynamics of the planning process itself, and the public's role in it.

Thus the way was paved for the various forest sector initiatives introduced by the new NDP government after its success at the polls in the fall of 1991. These programs, including the Commission on Resources and the Environment (CORE), the Forest Practices Code and others, were primarily exercises in pure process. Early in 1992, a few months after the election, a report was released that had a significant effect on the way these programs were received.

Tripp Biological Consultants had been hired by the government to prepare an evaluation of twenty-one logging operations on Vancouver Island to determine how well they complied with Coastal Fisheries/Forestry Guidelines. The report documented negative impacts on fish-bearing streams at almost every logging site examined. For proponents of IRM it was a timely demonstration of the need for more rigidly enforced regulations. The forest companies, which had for some time proclaimed their concern for non-timber forest values, were deeply chagrined. They found themselves in a poor position to oppose the new government's forest sector initiatives with any credibility.

The idea of a forest practices code, a set of written standards against which industrial activities could be judged, first emerged in the late 1980s. The forest industry set up a land use task force through the Council of Forest Industries to frame a code, as did environmentalists and other critics of industry, in a group known as the Tin Wis Coalition. Various proposals were made to the Forest Resources Commission. In its report, the Commission stated:

> There is a need for a single code that governs all the elements of good forest management, such as:
>
> - the choice of clearcutting or selection logging as a harvesting system within the silvicultural prescription.
> - the size, distribution and timing of clear cuts.

- approaches to be taken during harvesting to consider a wide range of other values, such as water in community watersheds, estuaries, streams and lakes, wildlife habitat and migration corridors; areas of high visual or recreational value; special soils or soil cover areas; slope instability.

The report went on to recommend that "a single, all-encompassing code of forest practices be established through the introduction of a Forest Practices Act."[10]

Seeing the writing on the wall during the last weeks of the Social Credit administration, Forest Minister Claude Richmond instructed the Forest Service to begin drawing up a code. By the time the NDP came into power, the bureaucrats were hard at work on the task. Early in its mandate the NDP was confronted with the spectre of a European boycott of BC forest products and eagerly embraced the FPC as a means of assuring European customers that logging was being done responsibly in BC. Fearing a boycott, industry leaders readily went along with this argument. In November 1991, Premier Mike Harcourt presented a discussion paper begun by the Forest Service under the Socreds. It had now become a full-blown set of laws, rules, regulations, penalties and provisions for its own administrative apparatus. Included were explicit descriptions of how to regulate logging, forest management and grazing practices to protect wildlife habitat, fish-bearing streams and other non-timber forest resources. The Ministry of Forests was to share the approval of development plans with the Ministry of Environment.

In several stages, the bureaucrats produced a series of documents outlining the FPC. First was a general piece of legislation, the Forest Practices Code of BC Act. This was followed by a set of regulations describing the basic practices that would prevail throughout the province. Next came a set of standards, one provincial in scope and the other specific to different regions of the province. Several dozen field guides followed, outlining detailed forest practice requirements, procedures, processes and expected results. A Forest Practices Board was set up to oversee mandatory audits of industrial operations, and an Appeals Commission to provide a review of decisions. An Enforcement Branch was set up within the Forest Service with an initial allocation of 200 staff. Maximum fines were raised from $2,000 per infraction to $1 million.

When the full scope of this exercise was revealed, the industry went into a state of shock. Initially it had gone along with the idea of an FPC, partly to satisfy its critics and, more important, to help forestall a European boycott. When the Forest Service produced the Forest Practices Act, which was fairly broad in scope, industry response was generally favourable. Even the regulations caused no great alarm. By the time the standards were released, however, consternation was mounting. It was beginning to look like the government was going to instruct industry, in excruciating detail, how to operate in the woods. By the time the field guides appeared, the industry knew it was facing a regulatory nightmare. Still, there was little protest. The boycott had been abandoned, the environmentalists appeared pacified and the media was shouting the virtues of the new code from the rooftops. The war in the woods, the pundits proclaimed, was over.

The code went into effect in June 1995. Within weeks the industry was up against regulatory gridlock as the thousands of approvals required before logging could begin plugged the bureaucratic pipelines. Within a few months, logging operations were being curtailed and postponed. Instead of being presented with a set of standards by which the results of its work in the woods could be measured, the industry was being told precisely how to operate in Crown forests. And then, to add to the woes, the government announced the FPC would also apply to private managed forest land.

There was strong resistance to this last measure from the major forest corporations that owned private timber land, as well as from small landowners. Their objections eventually led to development of a Private Forest Land Regulation, which attempted to achieve the same results as the FPC by focussing on the expected results of forest practices, rather than imposing a detailed set of regulations designed to achieve those results. When the regulation was completed in 1999, there was a widespread expectation it would serve as a model for revising the FPC itself.

The evolution of the concepts of multiple use, integrated resource management and forest practices codes over the past fifty years underline some of the basic difficulties and flaws in forest management policies in BC. Human use of forests typically entails a very narrow range of interests, primarily timber production. But each forest is a complex ecosystem of other life forms, physical attributes such as water and minerals, and more nebulous characteristics such as climatic influences, oxygen generation and human aesthetic values. It is extremely difficult to extract economic benefits from a forest without affecting its other aspects.

BC has retained state ownership of the forests and adopted a complex mass of laws, rules and regulations permitting the economic use of forests in certain narrowly defined ways. Onto this mass are piled more laws, rules and regulations in an attempt to protect other forest values and satisfy the desires of a growing number of interest groups. We call this activity multiple use or integrated resource management.

But true integrated management, or "sustainable forest management," as some now call it, would have forest landowners or licensees engaged actively in the stewardship of all the values in the forests under their jurisdiction. A forest company would be engaged in raising timber, along with fish, wildlife and all the other plant and animal species indigenous to the area. It would be concerned with the quality and quantity of water that flowed into and out of the forest, as well as the forest's effect on air quality and climatic conditions. It would not engage in these activities because a provincial enforcement officer armed with $1-million penalties was peering over its corporate shoulder, but because it made good financial sense to do so.

By the late 1990s there were still no incentives for people who work in BC forests to be concerned with all forest values. Under our system, forest companies have been concerned primarily with harvesting timber. In some cases they are also concerned with growing timber. They are not, in coastal BC for instance, concerned with sustaining and enhancing salmon stocks in the creeks and rivers that run through the forests they own or lease. In fact, such integrated management activity is prohibited by law. There is no economic or other link between the forest and the salmon, or the deer or other wildlife species sought by hunters and trappers. When it comes to non-game species, integration is even more problematic. And when it comes to northern spotted owls, it becomes downright difficult.

Under this system there has been no incentive for any forest manager to enhance spotted owl populations, though there are penalties for harming them. In other words, there are sticks, but no carrots. Our approach, including the new measures in the Forest Practices Code, has been to devise bigger and better sticks. The biggest stick of all is simply to withdraw large areas of forest land from the commercial forest land base. This may or may not sustain the spotted owl population of the province, and it has very little to do with integrating the management of timber and spotted owls. If history provides any indication of where this approach will take us, we can expect to end up with a smaller timber industry, a larger spotted owl bureaucracy and a declining population of spotted owls.

The existing forest management system is built on the premise that an educated technocratic elite, specially trained and backed by the authority of the state, can understand the ecological complexities of forest land, but forest workers and owners cannot—and if they can, they will not act in the interest of the forest as a whole. Another premise is that state agencies and their employees always work in the public interest, while private entities and their employees always work against the public interest. The history of forest resource management in BC tells us that none of these notions is true. The cumbersome system in place in the province has consistently failed to meet our objectives, largely because these objectives have never been defined before the means of achieving them are adopted. We must look elsewhere for a resolution of the resource use conflicts that plague us.

Looking Ahead 11

The British Columbia forest industry is in trouble, at times verging on collapse.

The industry has lost its pre-eminence in the minds of BC residents. Some foresee the industry being replaced with glamorous new economic activities based on advanced technology, others with increased tourism. Investors have all but abandoned the industry, declining to provide the funds needed to keep it functioning and competitive. Share values have plummeted; a few relatively well-off companies, or companies based outside BC, are poised to take over those that falter, which will escalate further the monopolistic character of the industry's structure. Some companies are verging on bankruptcy, with the odd one allowed to fall over the edge, particularly if it does not operate in a prominent provincial cabinet member's riding. Thousands of workers have recently lost their jobs and the survivors are engaged in a desperate and debilitating struggle keep the companies that employ them afloat.

The origins of this affliction go deeper than the market decline associated with the Asian financial crisis of the 1990s and other cyclical fluctuations. Overall, markets for forest products have been healthy, as is demonstrated by the financial strength of the industry outside the province. Nor is the softwood lumber agreement with the United States to be blamed. It is a factor but is more a symptom of the industry's malaise than a cause. This same agreement applies to other Canadian lumber producers that are doing very well in the US market.

It is tempting for some in the industry to blame the environmental movement for the plethora of rules and regulations that now restrict industry activity and its working area. This view is reinforced by those environmentalists who consider the condition of the industry a cause for celebration and an opportunity to restrain it even further. The industry they have fought for the past twenty-five years is in a sad and sorry state.

For a long time the forest industry ran BC. It was arrogant, demanding, and paid no attention to its critics. Many who worked in it have acted as if they had a God-given right to treat the forests as they wished, with no need to answer to anyone. Some still do. By and large the environmental movement has been good for the forest industry, which had become too powerful to correct itself. It had come under the control of its most rapacious members, to the dismay of many if not most of the other people who worked in it, and it was brought to its senses.

It is always tempting to blame the industry's problems on the government of the day, even more so the NDP regimes of Mike Harcourt and Glen Clark. Both governments were incredibly inept at dealing with industry problems. To its credit the Harcourt government did try to solve problems created by previous governments, which had allowed the industry to get too far out on a precarious limb for the sake of quick, easy profits. Unfortunately the Clark government was merely a mirror image of various Social Credit administrations, sacrificing the welfare of the industry to the benefit of its friends in organized labour, just as the Bill Bennett government, for instance, sacrificed it to the large forest corporations. Perhaps the greatest fault of the two NDP governments in their handling of the forest industry was that they were no better than many governments before them: they appeared to be self-interested, unable or unwilling to define objectives or see the larger picture, unwilling and unable to grapple with the underlying condition of the industry.

Another of British Columbia's great natural resources—salmon—is in a similar state. It too has the potential to be infinitely renewable and, perhaps not coincidentally, the industry based upon it is also in a state of ruin. Yet it is administered by an entirely different level of government. This would suggest that the problems afflicting BC's resource industries cannot be blamed entirely on any particular government, but are caused by the heavy proprietary role of the state itself in its various administrations.

The real indicator of the forest industry's condition may be the predictions for the future of both inside and outside observers. In 1997 the total BC timber harvest was 68.6 million cubic metres, compared to 90.6 million in 1987—a decline of more than 25 percent. It is widely expected that this harvest level will decline even further, to as little as 50 million cubic metres, and remain there for the foreseeable future. Such pessimism can only come from a demoralized and despondent industry.

The forest industry has changed enormously since the early 1970s, and it needs to change even more. In attempting to shift its focus and approach, the industry has come up against constraints and limitations resulting from policies and structures devised for a previous era. If the industry is to realize its potential and become the kind of industry the people of BC appear to expect of it—that is, the economic foundation of the provincial economy and the responsible steward of the province's forests—more fundamental structural changes are required.

Even in what may well be its darkest hour, the economic and social opportunities available to the BC forest industry, and therefore to all British Columbians, are brighter than they have ever been. Above and beyond current market cycles, world demand for forest products is growing steadily. Every year an additional 60 million cubic metres of industrial timber is needed, about the equivalent of BC's current annual output. Not only does this demand assure a market for timber, it also keeps prices up, which makes possible improved methods of forest management. Widespread awareness of the role forests play in global ecology has engendered in consumers a willingness to pay higher prices; they no longer expect forest products to be plentiful and inexpensive.

The forest industry's ability to obtain timber without destroying or irreparably damaging the forest is developing rapidly. Forest ecology has become a respected and mature science. Since the 1980s, new harvesting machines have evolved that are designed to move lightly through the forest without damaging it. In the hands of a new generation of forest workers, whose attention is focussed less on the timber they are removing and more on the forest they are leaving behind, this new technology can help transform industrial use of the forest.

This is not a time to narrow the scope of the BC forest industry. Nor is it a time, as some of the old industry war horses would prefer, to "get back to normal." What some industry people still look upon as the "good old days" are gone forever. The industry cannot concede a few more parks, agree to some operational modifications, then return to the time-honoured business of knocking them down and dragging them out. Fundamental changes are needed. And before we identify them, we need to decide what social objectives we want to pursue. Our forest policies should be conceived as strategies to achieve the broad goals we as a society decide on. The forest industry should serve its citizens, employees and shareholders—not the other way around.

This means we need to give some thought to the future as it might be fifty or a hundred years hence—not to create a plan or a detailed road map, but to imagine the kind of society we want for our children and our grandchildren. In BC we are more accustomed to focussing on the day-to-day tasks of developing and building a social and economic infrastructure. But the people who built the institutions and legal structures that now shape our lives were pioneers, living and working in a frontier society. For them it was enough to learn to live in the new world, with all its unfamiliar resources, and to keep up with technological developments. It must have seemed as though the forests would never be exhausted and, if they were, that more forests would be found. In our minds and dreams, we descendants have settled down, and we realize our future will consist of what we make of what we have, where we are. But many of our policies and institutions, including those pertaining to forests, still reflect old ways of thinking.

In order to thrive over the next few decades, the BC forest industry needs the full and enthusiastic support of the people of BC. It is not enough for various people with a direct interest in forests—"stakeholders," in the current jargon—to reach agreement on how the forests should be used. Objectives and policies that are widely accepted and supported are needed within and outside the industry.

One of the basic objectives of BC forest policy should be to achieve a high level of social diversity. Like most North Americans, British Columbians are a pluralistic people. In commissions of inquiry, elections and by other means, they have repeatedly made it clear that they do not want to live in a monolithic society. They are not enamoured of working only in large corporations, being represented by large unions, and regulated by large government agencies. They are not opposed to the existence of these large bodies, but they favour a healthy mix of large, small and mid-sized firms, family firms and individual proprietorships, union and non-union workplaces, and local government agencies along with centralized ones. Not many British Columbians desire a forest sector dominated by a few large corporations, two or three unions, a few multinational environmental organizations and a massive state bureaucracy. Yet that is precisely the type of structure BC's forest policies have created. Just as a healthy forest requires a diversity of species, so does a healthy society requires diversity in its social and economic institutions.

Another priority should be to rejuvenate and strengthen the province's rural communities. Since World War II there has been a semi-official administrative policy to transfer control of the provincial economy to large urban centres, particularly Vancouver and Victoria. It is always more troublesome and expensive to provide government services to far-flung rural communities than to large urban centres, where concentrated use of schools, hospitals, mail service, hydro lines, public transit and other services reduces per capita costs. Similarly, IWA Canada has found it far easier to organize and serve urban sawmill workers than loggers living in small, scattered rural communities. And in some parts of the province the large corporations that came to dominate the industry preferred employees living in logging camps to those living nearby the area of their operations; they were a captive workforce, single-mindedly undertaking the task of harvesting timber, and not distracted by personal, family or community concerns.

This kind of thinking—that people should cost as little and produce as much as possible—permeated policy-making in the past and was partly responsible for concentrating populations in urban centres over the past fifty years. There was no economic need for an established rural population when the primary activity was resource extraction. Knowledge of local conditions over time was not particularly important. Now that the job of managing forests has changed to one of resource stewardship, we do need to have an indigenous population, settled and stable, willing and able to undertake forestry work. For two or three decades we have been trying to accomplish the initial phase of this task—reforestation—with a migrant workforce, and while we have been fairly successful at it, we have not retained in those forests the knowledge and experience of that workforce. It has gone elsewhere, in most cases right out of the industry. If BC is going to get serious about developing its capability as resource stewards, the hinterlands need to be settled.

To bureaucrats, technocrats and other urban adherents to the gospel of efficiency, this proposition is probably quaint and naive. But given the opportunity, many British Columbians would choose to live and work in a rural forest environment. Clearly it is economically feasible: the task is a long-term one, and we know from examining other forest economies that forest management can be a productive, profitable activity. The old problems associated with the hardships of isolation have been overcome by modern transportation and communications technology, which make virtually the entire province habitable. There is no sound reason to encourage most of the citizens of BC to live in large urban centres. In fact, one of the major challenges our society faces is to slow the growth of big cities and to nurture a more diverse and decentralized population. We need to keep our large cities, but we also need smaller urban centres and rural villages.

BC's large urban centres stand to gain enormously from renewal and expansion of rural economies. Every dollar produced outside the cities will generate at least another dollar in the cities. This is not an argument against cities and the kind of modernistic, high-tech life they provide for their residents. Rather, it is a call to restore the balance between urban centres and the rural areas that provide much of the wealth. The objective is not to provide an alternative to high-tech visions of the future, but to provide a firm foundation of a resource-based economy upon which new, less certain, exciting endeavours can be built. One of the most highly regarded and successful high-tech companies in BC over the past two or three decades has been MacDonald Dettwiler, which is known around the world for its satellite technology. This technology was not created in an urban vacuum. It was developed initially for use by resource industries, primarily forestry and mining. A prosperous high-tech economic sector should not be built not as an alternative to a resource-based economy, but as a part of it.

Another objective of our forest policies must be to maintain and enhance non-industrial forest values. The economic use of forests, for timber harvesting, tourism, recreation or any other activity, cannot be allowed to diminish or destroy the forest. We must be careful not to kill the goose that lays the golden eggs. Much of the economic, aesthetic, spiritual and ecological wealth of BC is derived from its forests. No activity can be permitted that destroys or permanently diminishes this wealth. To maintain the health of our forests we have to understand that individually and collectively we are creatures of the forest. Our actions, whether we are listening rapturously to the singing of the trees or clearcutting an entire watershed, are natural. But some of our actions are not prudent or intelligent, and those are the ones we have to work on.

What we do in the forest is a social issue. The way we go about this task shapes the character of our society. Until now we have chosen to define our forest practices by adopting an ever more restrictive set of laws, regulations, penalties and other devices of command and control. This is very expensive, it is not successful, it is socially divisive, and the more we do it, the closer we move toward an authoritarian society.

There are other means of achieving the social objective of sustaining the forest. They include education and incentives. Education, unlike propaganda, seeks to enlighten and inform. It is based on the assumption that the more the people of BC understand forests, the less likely they will be to destroy them. A society uses incentives to reward desirable behaviour, and penalties to punish undesirable behaviour. Our forest policies contain a mixture of incentives and penalties. The problem is that they were adopted at different times to achieve different objectives.

Existing incentives are primarily economic, rewarding efficient logging. They were adopted in an era when no one particularly cared about the ecological consequences of logging. Fifty years ago, when the government decided to limit the amount logged, it offered longer-term leases in the mistaken belief they would encourage forest companies to grow their future timber supplies. After an initial spurt of silvicultural enthusiasm, forest companies failed to invest heavily in their leased forests, so the government began imposing a series of penalties, which became increasingly numerous and detailed and which culminated in the Forest Practices Code. There are still no incentives for companies to invest in and practise good forest management; there are only penalties for not following the rules. Even those who possess relatively long-term, area-based Tree Farm Licences and Woodlot Licences have no confidence they will realize the benefits of any silvicultural investments they make.

Education and incentives are not alternatives to regulations and penalties. In most situations they work best in tandem. If we think of incentives as carrots and regulations as sticks, we can see that education provides the knowledge people require to engage in forestry, and how to tell a carrot from a stick. People are usually much happier and more productive striving for carrots than being driven by sticks, although it is sometimes necessary to use a stick to draw their attention to the carrot. In BC, a good many of our sticks could be transformed into carrots, and a few could be retained for slow learners and social deviants. The idea should be abandoned that our citizens cannot be trusted to act in their own best interests, and that people require a small army of bureaucrats, experts, enforcement officers and other authority figures watching over them in order to act responsibly in the province's forests.

To achieve these social objectives, BC needs to adopt a forest sector strategy that reinforces them. Such a strategy would have several overlapping components. The first is to complete the conversion of the timber harvesting and processing industry into a forest management and forest products industry. This involves some subtle but fundamental shifts in emphasis—from thinking of ourselves as loggers, sawmillers and pulpworkers, to thinking of ourselves as forest managers and manufacturers of forest products. If logging is comparable to the harvesting of wheat or barley, forest management can be seen as the equivalent of farming. Farmers do not consider themselves primarily as harvesters of grain; that is part of their work, as is planting seeds, tending crops, conserving soil and so on. From this perspective, silviculture is an opportunity, not an obligation. It is a critical part of the business, not a burdensome necessity. Reforestation is the beginning of a new forest cycle, not the final stage of a timber-harvesting operation. Until we think in these terms, we will not be managing our forests.

Second, we need to think of ourselves as manufacturers of products rather than producers of commodities. Two spruce–pine–fir studs, for instance, may be identical in their physical properties. The difference between them lies in the minds of the people who made them and the consumers who buy them. The commodity stud is an undifferentiated consequence of a mass-production manufacturing process. The stud-as-a-product is different because at some point during the manufacturing process it was decided the most valuable product that could be made from the log was an s–p–f stud. Other logs would end up as violins, or furniture, or sheets of plywood, but not this one. The other difference is that the second stud is more valuable than the first. It was made into a stud because it met the standards required of studs, and was sold with others like it to a consumer who could be confident of the quality of the products he was buying. A bundle of commodity studs contains good studs, poor ones and every grade in between, because the process turns every log into a stud. The consumer will not have the same degree of confidence in what he is buying, and will not be willing to pay as much as for a bundle of studs-as-product. The difference in value between the two studs is not very significant, but when the value of the violins, furniture and plywood are factored in, the added returns become substantial.

This distinction should not be confused with "value-added" manufacturing. While it is an improvement on previous manufacturing practices, the value-added approach is often merely an extension of commodity production. It seeks to produce more valuable commodities, for reasons that often have little to do with the qualities of the thing produced. For instance, a recent practice employed in BC commodity stud mills was to drill a hole through each stud. When these studs are used in construction, electrical wires can be threaded through the holes, saving the builder the work of drilling his own holes. The quality of these studs has not been improved in the manufacturing process, nor has the value of the mill's output been raised significantly. The process may have created another job or two, and permitted a small increase in price, adding value to the studs. In this case, the purpose of drilling the hole was to enable the pre-drilled studs to qualify as "value-added products" and avoid US tariffs on commodity lumber. This is a legitimate, clever ploy in the context of our overly regulated system, but it should not be confused with a strategy of optimizing our manufacturing processes and making the best use of the available timber supply.

If we reorganize our forest sector to maximize productivity of our forests and take the fullest advantage of our timber supply, we will enjoy greater economic benefits without destroying or diminishing our forests. Some claim we can triple or quadruple the economic benefits we obtain from the forests. The fact is we know too little about managing our forests to calculate potential gains accurately, but looking elsewhere may give us an idea of what to expect.

Finland, with a population of 5 million people, provides a useful comparison. The entire country of Finland lies at latitudes higher than BC's northern border, which means growing conditions for trees there are less favourable than ours. The Finns began the intensive management of their forests after World War II, about the same time we took up the idea of maintaining a permanent forest economy. Like us, they are located near a large foreign market, and they sell most of their products to other Europeans. Beyond these similarities, the two forest sectors are quite different. Two-thirds of Finland's forests are privately owned, in holdings that average 37 hectares; 80 percent of all holdings are smaller than 50 hectares in size. Yet their forest product companies, which have to compete in international markets, are much larger and stronger than ours. The Finnish logging and milling industries are more mechanized than ours, their workers more productive, better educated and higher paid. Several decades ago, the people of Finland adopted a set of strategies that now provides them with economic benefits far in excess of ours.

FOREST ECONOMIES IN 1994

	Finland	BC
Forest area outside parks	26,300,000 ha	37,000,000 ha
Annual forest growth	81,000,000 m^3	Unknown
Annual timber harvest	49,000,000 m^3	75,600,000 m^3
Direct employment	99,000	95,900
Jobs/ 1000 m^3	2.02	1.27
Jobs/ 1000 ha of forest	3.76	2.59

Sources: *Annual Ring*, Finnish Forestry Association, 1996; *The Forest Industry in British Columbia*, Price Waterhouse, 1995.

These figures indicate that on a per-unit basis, the Finns obtain about 50 percent more direct jobs than we do. But the figures do not show the production values of the two jurisdictions, or the number of indirect jobs. Finnish companies produce a much higher proportion of manufactured products and fewer commodities than BC companies. There are also more companies, engineers, forest consultants, equipment manufacturers and the like, serving the global forest industry in Finland than in BC. The Finns have used the knowledge they acquired in their own forests to become the world's primary producers of forest machinery and technical expertise. These accomplishments are well within the reach of British Columbians. We could expect a 50 percent increase in forest sector employment, with higher wages and even greater gains in value, if we adopted clear social and economic objectives and implemented the strategies required to achieve them.

As well as timber, there are other economic benefits to be derived from BC's forests. In 1994, Finnish forest owners harvested almost $35 million (Can) worth of wild berries, and more than $50 million worth of venison from managed reindeer herds. In BC the potential for forest foods, fish, wildlife and other non-timber values is probably even higher. Some of today's potential did not even exist in the past. For example, unpopulated coastal watersheds could be managed to produce pure, fresh water, among other things. Dozens of such watersheds are accessible to deep-sea tankers, which could transport water to foreign markets. The gathering of wild mushrooms has become a large, lucrative business; this could be expanded enormously by managing forests to enhance mushroom production.

What is needed is a structural context in which whole-forest management, as opposed to single-value management, can be practised. Forests consist of far more than timber and they should be managed as coherent entities. Forest policy in BC has been directed at providing such coherence or integration of management through centralized government agencies that oversee the activities of various tenure holders. In most cases this approach has failed. Instead we need multiple-resource management tenures and more comprehensive rights included in ownership. Timber and grazing leases could be combined, and management tenures could include management of fish and wildlife resources.

BC cannot attain the social objectives of rural rejuvenation, enhancement of non-industrial forest values and social diversification by downsizing the forest economy, as has been done through the late 1990s. Instead, we must develop the skills and resources needed to care for our forests. The history of the twentieth century demonstrates that economically poor societies destroy their forests; prosperous ones do not. The rate of deforestation in developing countries between 1980 and 1990 was twenty times that of industrialized nations.[1]

Reconfiguring the BC forest sector cannot be accomplished overnight. First, comprehensive forest management skills need to be developed. Currently a large number of professionals and technicians work for government and industry as skilled administrators. They are experienced in the bureaucracy of silviculture and amply able in filling out forms, obtaining permits, measuring timber volumes and land areas, and overseeing contract workers. Unfortunately, only a few of them have had much direct experience in managing a forest. And fewer still have managed a particular forest over a lengthy period. BC foresters must be able to get into the forest, managing forests. Some of the silvicultural workers who do most of the actual work of planting, spacing, fertilizing, pruning and harvesting trees have acquired good skills in certain specialized tasks. But they spend little time in a particular forest, and rarely work in the same forest site more than once. Silviculture is a site-specific undertaking, and long-term knowledge and experience of specific forests is required. It will take time to develop the skills needed to increase forest yields and values.

Second, it is important to maintain the economic stability of the existing industry and of BC's forest-dependent communities. Disruptions and uncertainties created by changes introduced over the past few years have cost the industry vast amounts of money, have generated panic and anger among workers whose jobs have been threatened, and have shaken the economic foundations of most forest communities. A transition period of ten to fifteen years is needed, during which a fundamental restructuring of the industry can take place without damaging the industrial operations already in existence. Policy must be stabilized during this transition period: change cannot be managed with each successive government directing the industry down a different road.

Third, it is necessary to honour contractual obligations already in place. Some critics advocate a wholesale cancelling of existing leases and licences, and the reallocation of tenures. There is a terrific penalty to pay for taking this approach. If government cannot be relied on to honour its commitments, but simply changes the rules to suit its own agenda, then no one will trust government enough to undertake long-term investments of money and energy.

Fourth, the restructuring of the industry needs to be done in concert with settlement of Native land claims. It has been argued that nothing can be done about reforming the forest tenure system until the land claims issue is settled, but this is more of an excuse for not making change than a reason for being unable to do so. The issues of land claims and forest tenures are related and need to be resolved concurrently.

Change—even sweeping change—is possible. Few people engaged in the forest industry would argue change is not needed. But change cannot be thrust upon the industry by government. Choices and options can and should be offered to those who already possess rights to use the forest. Discussion, involvement and negotiation are the civil means to a restructuring, not confiscation and arbitrary pronouncements.

The most fundamental change needed is overhaul of the forest tenure system. To reflect the social and economic composition of the province, we need a diverse range of private forest holdings, corporate ownerships and public ownerships. For lack of better criteria, we could aim for ownership ratios common in other developed forest countries—one-third of the land in corporate ownership, one-third in smaller private holdings and one-third in collective ownership, primarily some form of "community forest." Community forests should be owned and operated by regional, municipal and Native governments. Tenure strategy should be based on two principles: a substantial increase in private and corporate forest ownership, and the transfer of the remaining provincially owned forests to local, regional and Native governments. The success of municipally managed forests at Mission and Revelstoke indicate the great potential of this approach. Resource planning and regulation should also be performed at the community level. The objective is to democratize resource use planning through a genuine decentralization, or devolution of power and control to responsible government at the community level.

The methods we use to diversify forest ownership and tenure are important because they will affect forest management. One method is outright sales, based on the assumption that if individuals and firms invest money in land as a business venture, they will be unlikely to abuse their investment. This method is well established in our society and would serve in some cases. It would also provide the Crown with revenue. However, if this were the only way to acquire forest land, it would transfer a great deal of wealth from the private sector to government. It may be better to devise some means of transferring title so this money could be invested in the forests rather than turned over to government. In the long run, the public treasury has more to gain in revenues through taxes extracted from a vibrant forest industry than it does from a one-time sale of forest land.

There is also a question of fairness. Cash sales would limit buying opportunities to those with money or the ability to borrow it, who may not be the best or the only good forest managers. A system of "sweat equity"—earning title to the land—could also be implemented. Area-based leases such as existing Woodlot Licences or Tree Farm Licences could be granted to individuals, families or companies with the proviso that the licensees could earn title to the land only if they demonstrated their ability as responsible and capable forest managers over a period of ten to fifteen years. Historically this method has proven successful; it is similar to the homesteading system by which much of the prairies and parts of BC were settled by newcomers early in this century. There are undoubtedly many other means of transferring title, and BC could adopt a flexible combination of them.

The question of how to transfer title is complicated somewhat by existing area-based tenures. Of the 50 million hectares of productive forest land in the province, about 4.7 million hectares, or less than 10 percent, lies within Tree Farm Licences and Woodlot Licences. Most of this land could be managed to optimize its productive capability if the proper incentives were in place, and in the case of some TFLs, licensees will be interested in this option. Other companies look upon TFLs as harvesting tenures and may have no interest in long-term forest management. The trick in this case is to separate the wheat from the chaff by offering ownership, long-term leases or other tenure arrangements to permit the interested licensees to get into the business of managing forests, and allow the others to phase themselves out of these licences as they are harvested, turning over the second-growth forests to someone else for long-term management.

Tree Farm Licences were not devised primarily as forest management units but as timber harvesting operations, and there has been a great deal of corporate consolidation since the licences were issued. Consequently the configuration of some TFLs may not be conducive to management—they may be too big or too scattered. In 1997 a bond-rating agency determined that MacMillan Bloedel's Crown tenures, mostly Tree Farm Licences, were a financial liability. In cases such as this, and perhaps for licensees who do not want to engage in forest management, we need a means of transferring title or tenures to new corporate entities owned by the people who now work in them—workers, contractors, managers and so on. In most cases the local TFL workforce has a long-term interest in the forest covered by the licence; consequently it has established some rights over the disposition of the forest. For TFLs we need a clear articulation of our objectives and a variety of tactics for negotiating a restructuring.

The situation is quite different in the Timber Supply Areas, where only about 250,000 hectares is committed to area-based tenures in Woodlot Licences and the new community forest tenures. This leaves almost 41 million hectares of productive forest land in the TSAs, all of which can eventually be converted to private and corporate ownership and community forests. Existing contractual commitments in the TSAs are for annual volumes of timber under such tenures as fifteen-year Forest Licences. In almost every TSA there is sufficient old-growth timber left to supply these licensees with their allocated timber supply well beyond the term of their licence, some for another fifty or sixty years. For the most part, only second-growth forests should be converted to private ownership, and as old-growth forests are logged they too can be converted. This way there is no need to deprive established operations of their timber supplies while phasing in new forms of tenure.

This type of wide-ranging tenure reform raises numerous ancillary issues. For instance, what about variations in existing timber values on different parcels of land? A tract covered by a fifty-year-old stand of timber is worth a lot more than a tract of recently reforested land. One method would be to treat the timber on a piece of land converted to private ownership separate from the land itself. It could be appraised and purchased separately, either at the time of land acquisition in a lump sum payment, or when it is eventually harvested, in the form of stumpage.

What would prevent the new owners from liquidating the timber and pulling up stakes as soon as they obtain clear title? For one thing, self-interest: growing timber is one of the best investments available, with longer-term security. People will tend to realize this and behave accordingly. To reinforce this inclination, BC could adapt one or more European methods of ensuring that forest land is not mistreated. A common approach is to require landowners to post a bond before harvesting their timber. Once they have reforested or restored their forest to a productive state, they redeem their bond. If they do not perform as required, the local forest authority uses the bond to have the work done. This and other methods have been tried in other parts of the world, and variations could be adopted for BC.

How would non-timber forest values be protected? By laws and regulations that would spell out the standards of performance expected and the penalties for failing to achieve these standards. A set of Forest Laws would define the kinds of things that can and cannot be done with forests, as in most European countries. Basically forest land would be treated in a different way than other types of land, recognizing that it has values extending beyond its boundaries and the immediate interests of its owners. For instance, owners of forest land would be expected to maintain wildlife habitat. Incentives could be provided through the tax system to encourage this, backed up with regulations should they persist in the folly of acting against their own financial interests.

What about the public's current right of access to Crown land? Much of this access could be maintained. Transferring forest land to private ownership does not mean conferring on it the property rights of other types of private land: the public already has access to private forest land for certain clearly defined purposes such as picnicking, hiking, etc., but not for others, such as cutting trees, hunting or camping.

The Ministry of Forests takes the position that under its yield regulation system, additional area-based tenures cannot be allocated in the TSAs because the timber supply is fully committed. The simple solution to this problem is to adopt new yield regulation systems, one for mature or old-growth forest land and another for immature or second-growth forests. In the new forests there will be no need to regulate the yield, as we do now. If the forests are well managed, the potential yield of individual forests will be realized; together these yields will meet an optimum sustained yield.

Regulation of the yield in the remaining old-growth forests should proceed according to objectives determined by the community in which the forest lies. In some cases it will be prudent to parcel out old growth for several decades to provide economic stability. In others it may be desirable to log decadent or diseased old growth more quickly in order to replace it with healthier young forests that will provide enhanced future benefits. The important point is that within broad provincial goals, decisions about how to regulate timber harvesting levels in these forests should be made by people in the communities directly affected.

It is important to recognize that harvesting remaining old-growth forests designated for commercial use (as opposed to those that will be protected) is only a short-term action that will provide a transition to the longer-term business of managing new forests. At present there are about 17 million hectares of productive second-growth forest in the TSAs and TFLs. Between a third and a half of it is old enough to provide sufficient revenue to underwrite the costs of more intense management. Most of it—14 million hectares—is outside area-based tenures and is well suited for allocation to individuals, forest companies and communities. Assuming there are at least 5 million hectares of such advanced second-growth forest, and assuming conservatively that every 1,000 hectares would support three workers, getting this land into the hands of people capable of managing it would directly employ up to 15,000 forest workers and probably twice as many more workers indirectly, almost all of them in hinterland communities. If these forests were managed using conventional silvicultural methods employed in Europe and the US, including salvaging mortality, thinning, improving stand composition and so on, they could produce at least 100 cubic metres of timber per hectare every twenty years. This level of management would provide an additional 25 million cubic metres of timber each year, which would directly employ some 10,000 to 15,000 millworkers and support even more indirect jobs. As more second-growth land begins producing useable timber, these numbers will grow. The results will not be realized overnight, but they are possible in the long run. They will never materialize if the existing structural arrangements do not change.

In addition to these older immature forests, there are about 9 million hectares of younger immature stands, and about 1.6 million hectares of poorly stocked land. To this area, annual clearcut harvesting in the TSAs adds more than 100,000 hectares. Most of this land will not produce revenue from its silvicultural operations until it is older, and intensive management of it can require a substantial financial investment. This land is suitable for allocation to larger firms and communities, which are in a better position than individuals and small operators to raise investment funds in financial markets. The goal should be to allocate the existing immature area over a specified period, say ten years, and as additional land is logged, it should be converted to an area-based tenure, convertible to private or community ownership. If logging companies working on volume-based licences were offered first rights to management tenures on the areas they log, they would have an incentive to improve their harvesting practices.

Full-scale management of the young immature and unstocked

area—about 10 million hectares—would require a full-time silvicultural workforce of perhaps 10,000 to 20,000 workers and annual investments of between $1 and $2 billion. The biggest problem here is not financial, it is a lack of skilled workers and a structural context in which they can function productively. Plenty of investment money is available to a suitably structured forest sector from pension funds, insurance companies and other agencies. But training and educating workers will take time, so BC needs to get on with the job.

The skills and education required to manage our forests competently are substantial. Until now, many loggers, millworkers and silvicultural workers have been drawn from the bottom of the labour pool. It does not take much skill or training to set chokers, plant a seedling or pull boards off the green chain. But in the forest industry that is evolving, these kinds of jobs will no longer exist. In some parts of the world there are no loggers at all working on the ground. Partly this is because it is difficult to find people willing to do such physically demanding work. Also, with the increased use of more sophisticated machinery, it has become much more dangerous for workers to function outside machines. And, in order to pay workers the wages they need, companies must supply them with the tools and machines they need to be productive. Inevitably this leads to increased mechanization of forest work involving the introduction of more sophisticated equipment. When a skidder cost $15,000 and was used only to move logs to the landing, with little regard for the forest that would follow, a skilled worker was not needed to operate it. Clearcutting a forest required only skill in falling trees and maintaining a chain saw. Today, the machines that work in managed forests cost half a million dollars each and run on complex computer technology. An operator must undergo a long training period, because he or she must be capable of making knowledgeable silvicultural judgements, running a sophisticated harvesting machine with skill and efficiency, and repairing the machine when it breaks down. People with these abilities are paid far better than old-time loggers, and they need training and education.

The manufacturing end of the industry also needs more highly skilled workers than those it has traditionally employed if the value of production is to be increased. For example, there are many proposals to add value to our forest products by manufacturing furniture from BC lumber. We certainly have the raw materials, and there is a large global furniture market. However, making marketable furniture is not simply a matter of hammering a few boards together into a table or a chair. The key to success in the furniture business is design. No one will buy poorly designed furniture. After World War II, the Scandinavians dominated the global furniture business by producing new, stylish designs that swept the market. British Columbians could follow a similar course, but first we will have to develop a superior design capability, which will take time and money. If furniture-making is to be included in BC's industrial strategy, one or two well-funded industrial design schools need to be established immediately; they will need to be maintained for some years before BC develops a pool of designers who can produce world-class furniture.

Another needed strategic change is to restructure forest revenue policies. As timber harvesting shifts from old-growth forests to managed second-growth forests, government should phase out its reliance on timber-based revenues such as stumpage and royalties, and look instead to corporate and personal income taxes, land taxes and other conventional tax devices. The objective should be to raise public revenues from forests after management expenditures and investments have been made. Timber-based charges lead to state exploitation of the forests, and BC has had a long history of that. A necessary part of a new forest sector strategy is a public revenue system that encourages forest management and provides incentives to reinvest forest earnings in the forest.

Other taxation policies should be adopted to ensure adequate funds are available for forest management. Forests provide good opportunities for those looking for secure long-term investments such as pension or insurance funds. The existence of secure private forest tenures might be sufficient to attract these funds; if not, tax incentives could be used to provide management of non-industrial forest values such as air quality or carbon containment.

If the free export of logs from managed second-growth forests were permitted, large amounts of revenue would accrue to owners of forest land, and tax policies could be used to encourage the reinvestment of those revenues in the forests where they were earned. Banning export of these logs in order to assist the domestic manufacturing sector merely transfers unearned revenues to mill owners and workers and fosters an uncompetitive mill sector. The creation of a large, land-based forest management industry would create a genuine log market in BC, which would be part of a global log market. Mills in BC would be able to obtain the grades and species of logs they need at market prices to provide products for the global marketplace. A manufacturing sector could develop that was based upon its superior productive capabilities, instead of on its access to a protected timber supply.

These few proposals for change suggest a direction in which forest policy might turn, but they are just a beginning. Forests and the forest industry are complex and far-reaching in BC, and many more issues, problems and details need serious attention and deep thought. As we have learned from the well-intentioned but poorly executed initiatives of recent provincial governments, half a change is often worse than no change at all: the changes still needed are fundamental and profound. They cannot be made one at a time, to solve each immediate problem. They require vision and foresight and they must be made in the context of clearly articulated social values.

The time is right for British Columbia to undertake the task of rethinking and reworking forest policy. Before now we have not had the knowledge or experience to take such a comprehensive approach to the forests: we were too caught up in using forests to build the foundations, the infrastructure and the institutions of our society to accumulate wisdom about the source of our wealth. Literally, we could not see the forest for the trees we cut down and sold to pay for our schools, hospitals, art galleries, highways and other necessities of contemporary existence.

In the last two decades of the millennium, the situation has finally changed. It was once a cliché that we could not expect to manage our forests as well as older civilizations because we lacked a "forest culture." Now British Columbians share a widespread and passionate interest in forests, as the forest industry learned recently at great cost. It is now time to tap the skill and knowledge forest workers have accumulated but have not had a chance to fully exercise.

In some ways, experiences over the past twenty years have brought us face to face with the possibility of a terrible loss. We could lose the livelihoods of thousands of forest workers; we could lose a cornerstone of our provincial economy; we could lose the forests themselves. We have realized how important our forests are, how essential they are to our sense of ourselves as people inhabiting the BC landscape. That knowledge lies at the core of our forest culture, and it will be a powerful tool for us as we explore new approaches to using our forests in the future.

Appendix

TABLE 1A
Regional Analysis (1994)

CARIBOO	
Total area	8.2 million ha.
Forested	6.5 million ha.
Mature	3.7 million ha.
Immature	2.2 million ha.
Unstocked	.4 million ha.
Unproductive	.2 million ha.
Parks & protected	.2 million ha.
Population	62,100
Forest sector employees	7,000
KAMLOOPS	
Total area	7.5 million ha.
Forested	5.2 million ha.
Mature	2.8 million ha.
Immature	1.9 million ha.
Unstocked	.3 million ha.
Unproductive	.2 million ha.
Parks & protected	.7 million ha.
Population	372,500
Forest sector employees	12,000
NELSON	
Total area	8.2 million ha.
Forested	4.9 million ha.
Mature	1.7 million ha.
Immature	2.3 million ha.
Unstocked	.3 million ha.
Unproductive	.6 million ha.
Parks & protected	1.9 million ha.
Population	149,400
Forest sector employees	9,000
PRINCE GEORGE	
Total area	31.8 million ha.
Forested	22.4 million ha.
Mature	9.5 million ha.
Immature	6.7 million ha.

Unstocked	2.0 million ha.
Unproductive	4.2 million ha.
Parks & protected	1.5 million ha.
Population	164,100
Forest sector employees	14,200

PRINCE RUPERT

Total area	25.7 million ha.
Forested	11.9 million ha.
Mature	7.0 million ha.
Immature	2.1 million ha.
Unstocked	.5 million ha.
Unproductive	2.3 million ha.
Parks & protected	2.5 million ha.
Population	85,000
Forest sector employees	6,500

VANCOUVER

Total area	13.8 million ha.
Forested	6.9 million ha.
Mature	3.7 million ha.
Immature	2.3 million ha.
Unstocked	.1 million ha.
Unproductive	.8 million ha.
Parks & protected	1.7 million ha.
Population	2,446,900
Forest sector employees	55,000

TOTAL BC

Total area	95.2 million ha.
Forested	57.8 million ha.
Mature	28.4 million ha.
Immature	17.5 million ha.
Unstocked	3.6 million ha.
Unproductive	8.2 million ha.
Parks & protected	8.5 million ha.
Population	3,271,029
Forest sector employees	103,700

Source: *Forest, Range & Recreation Resource Analysis* 1994 (Victoria: Ministry of Forests), pp. 5-11.

TABLE 1B
Tree Farm Licences

No.	Area	Productive	Mature	Immature	NSR	AAC (1997) m³
	———	———	hectares	———	———	
1	609,204	286,955	208,945	68,873	9,137	720,000
3	79,128	51,588	21,667	28,263	1,658	65,000
5	34,320	32,954	11,815	19,576	1,563	110,000
6	170,026	143,823	81,090	39,826	2,907	1,288,000
8	77,400	71,600	43,100	27,148	1,352	145,000
10	231,686	62,378	45,761	15,253	1,364	170,950
14	139,443	53,632	28,919	22,679	2,034	164,000
15	48,106	44,207	17,248	24,558	2,401	78,000
18	74,571	67,441	30,528	27,997	8,916	187,000
19	195,100	113,145	71,957	40,522	666	978,000
23	554,981	371,132	173,005	186,539	11,588	680,000
25	458,180	210,487	177,799	51,516	3,172	779,000
26	8,904	7,659	723	6,228	708	45,000
30	180,766	159,292	84,757	43,252	31,283	350,000
33	8,458	7,768	4,026	3,385	357	22,500
35	39,199	37,356	17,383	17,648	2,325	125,600
37	189,851	136,685	77,861	56,489	2,335	1,063,000
38	218,408	63,239	49,408	11,876	1,262	263,000
39	794,143	537,952	327,410	196,102	14,440	3,740,000
41	1,019,740	270,899	241,698	29,201	0	400,000
42	49,990	46,779	31,566	10,282	4,931	120,000
43	10,165	5,792	3,854	1,938	0	44,460
44	450,712	372,872	223,873	138,923	10,076	2,228,000
45	242,760	189,975	33,522	13,700	5,563	220,000
46	108,988	102,191	37,009	61,866	3,316	535,000
47	224,614	202,464	48,307	154,157	0	865,000
48	638,770	519,770	299,745	172,206	28,268	514,000
49	144,923	134,481	72,190	55,768	6,523	380,000
52	265,190	244,339	172,574	62,016	9,749	549,000
53	87,415	79,619	42,308	31,05	6,296	204,700
54	67,637	59,651	43,244	7,570	8,837	75,750
55	92,228	45,605	33,326	9,835	2,444	100,000
56	119,505	59,915	39,517	17,229	3,169	100,000
	7,725,291	4,720,690	2,803,431	1,705,924	191,091	17,424,960

Area numbers are valid as of February 14, 1995. AAC is the allowable annual cut assigned in 1997. Source: BC Ministry of Forests amd Ministry of Forests Timber Supply Review Results, January 1997.

TABLE 1C
Timber Supply Areas

Area	Prod.Area	Mature	Immature	NSR	AAC m^3
hectares					
CARIBOO REGION					
Quesnel					
1,711,861	1,481,211	850,508	541,698	86,264	2,340,000
Williams Lake					
4,861,715	3,447,706	2,075,807	1,165,008	204,379	3,807,000
100 Mile House					
1,220,579	1,019,575	578,485	379,786	60,427	1,326,000
NELSON REGION					
Arrow					
753,948	489,424	147,390	308,315	23,277	619,000
Boundary					
580,110	461,769	141,207	294,459	25,145	700,000
Cranbrook					
1,391,619	842,373	261,964	515,747	61,130	850,000
Golden					
920,167	318,184	143,171	149,909	22,780	540,000
Kootenay Lake					
1,144,803	681,317	254,029	397,175	24,921	700,000
Revelstoke					
503,851	210,022	140,915	58,831	10,120	230,000
Invermere					
1,017,021	582,272	262,537	284,271	34,507	591,500
KAMLOOPS REGION					
Okanagan					
2,173,877	1,606,059	810,629	687,655	97,527	2,615,000
Merritt					
1,115,914	893,203	537,883	306,223	42,993	1,454,250
Kamloops					
2,118,685	1,567,186	825,738	634,341	102,983	2,679,180
Lillooet					
1,123,819	605,949	456,969	122,856	23,547	643,500

PRINCE GEORGE REGION					
Dawson Creek					
2,277,753	1,661,425	884,541	614,735	79,104	1,733,033
Fort Nelson					
8,329,102	4,358,891	1,647,035	1,793,526	317,877	1,500,000
Fort St. John					
4,553,553	2,520,784	890,427	1,288,435	137,450	2,015,000
Prince George					
7,512,982	5,601,040	3,541,928	1,700,305	263,331	9,363,661
Mackenzie					
6,130,617	3,073,864	2,014,773	904,476	84,172	2,997,363
Robson Valley					
1,235,817	507,179	371,727	118,480	13,390	602,377
PRINCE RUPERT REGION					
Bulkley					
736,154	512,192	368,529	119,974	14,489	895,000
Lakes					
1,123,263	923,967	589,205	299,847	30,069	1,500,000
Cassiar					
13,413,227	3,838,809	2,631,683	1,001,105	90,114	400,000
Kalum					
2,163,771	880,958	683,662	118,585	39,654	464,000
Kispiox					
1,222,851	786,996	622,591	124,641	19,677	1,092,611
Morice					
1,498,698	1,032,390	693,010	292,889	37,999	1,985,815
North Coast					
1,947,973	739,131	681,724	48,502	8,545	600,000
Cranberry					
76,751	60,444	40,936	14,762	3,109	110,000
VANCOUVER REGION					
Arrowsmith					
994,610	168,727	69,570	95,502	3,623	400,000
Fraser					
1,165,178	614,475	265,779	321,081	26,165	1,550,000
Kingcome					
1,126,125	564,913	419,079	133,330	11,612	1,399,000

Mid Coast						
	2,214,984	814,322	710,790	91,185	11,819	1,000,000
Queen Charlotte						
	464,827	363,164	326,271	32,505	4,388	475,000
Soo						
	719,503	323,807	190,280	123,176	8,760	506,000
Strathcona						
	659,110	395,226	201,815	177,258	15,944	1,420,000
Sunshine Coast						
	1,121,130	514,198	193,323	305,027	15,736	1,140,000

Totals

81,325,948 4,4463,517 2,5525,882 1,5565,60057,882 53,430,290

Source: Areas from Inventory Branch, BC Ministry of Forests, February 1995. AAC from Ministry of Forests Timber Supply Review Results, January 1997.

TABLE 4A

Direct Employment in British Columbia
(number of full-time equivalent employees)

	1998	1997	1996
Lumber	21,000	23,000	23,700
Plywood	3,200	3,300	3,300
Market pulp	6,900	7,400	7,500
Uncoated groundwood papers	3,100	3,300	3,700
Logging—company	8,700	9,150	9,500
Logging—contractor	19,500	20,500	21,000
Value-added	12,500	13,000	13,000
Provincial government	4,400	5,000	5,000
Silviculture	4,600	4,600	4,600
Other operations	7,500	8,000	7,800
Total	91,400	97,250	99,100
Employees per 1,000 m^3 logs harvested			
Direct	1.4	1.4	1.3
Total BC	4.2	4.3	3.8

Source: Price Waterhouse, *The Forest Industry in British Columbia*, 1998.

TABLE 4B

Summary Financial Data 1992–1998

(Dollar amounts in millions)

	1998	1997	1996	1995	1994	1993	1992
Sales							
	$15,006	$16,205	$15,965	$17,676	$15,965	$13,837	$11,162
Direct Employment							
	91,400	97,250	99,100	97,500	95,900	92,200	92,200
Indirect Employment							
	182,800	195,000	198,000	195,000	192,000	184,000	184,000
Payments to Governments							
Federal							
	$133	$108	$91	$280	$413	$165	$9
Provincial and Municipal							
	1,904	2,336	2,272	2,506	2,414	1,439	1,006
Related to Direct Employees							
	1,648	1,738	1,892	1,873	1,740	1,560	1,420
Total							
	$3,685	$4,182	$4,255	$4,659	$4,567	$3,164	$2,435
Net Earnings (Loss)							
Lumber							
	(58)	$259	$392	$479	$1,136	$1,235	$182
Plywood							
	6	3	(2)	46	65	70	24
Market Pulp							
	(373)	(284)	(587)	393	(103)	(479)	(178)
Uncoated Groundwood Papers							
	46	(65)	90	112	(64)	(188)	(220)
Other Operations							
	43	(105)	(183)	250	326	(118)	(70)
Total from Operations							
	(336)	(192)	(150)	1,191			

Non-operating Items	(721)	60	(140)	89			
Total Net Earnings (Loss)	$(1,057)	$(132)	$(290)	$1,280	$1,360	$520	$(262)
Capital Expenditures	$729	$1,161	$1,392	$1,270	$1,064	$1,122	$907
Cash Flow from Operations	353	$1,130	$1,671	$2,083	$1,624	$537	$823
Short and Long Term Debt	$7,436	$6,661	$6,102	$6,521	$6,285	$6,146	$6,102
Capital Employed	$16,804	$17,702	$16,707	$17,303	$16,140	$14,469	$13,854
Return on Capital Employed (%)							
Lumber	0.7	5.2	8.6	10.5	21.5	28.8	9.0
Plywood	3.2	2.4	1.5	21.5	29.1	35.0	17.1
Market Pulp	(5.1)	(4.1)	(9.0)	10.0	0.5	(8.1)	(2.1)
Uncoated Groundwood Papers	3.2	(1.7)	5.3	5.8	(1.2)	(3.5)	(4.0)
Total	(3.9)	1.4	0.0	9.3	10.3	5.7	0.8
Return on Shareholders' Equity (%)	12.9)	(1.4)	(3.2)	13.7	15.6	7.1	(3.8)
Debt as % of Book Capitalization (%)	44	38	37	38	39	51	49
Log Harvest (millions of cubic metres)	64.8	68.6	75.2	76.5	75.6	79.2	74.0
Average Logging Costs on Crown Lands ($/m^3	79	88	83	79	69	54	48

Source: Price Waterhouse, *The Forest Industry in British Columbia*, 1998.

TABLE 4C
Log Volume Harvested
(million m³)

	1998	1997	1996
Coast			
Company	10.4	11.4	11.8
Contractors	8.5	10.9	11.1
	18.9	22.3	22.9
Interior			
Company	0.3	0.4	0.4
Contractors	45.6	45.9	51.9
	45.9	46.3	52.3
Total			
Company	10.7	11.8	12.2
Contractors	54.1	56.8	63.0
	64.8	68.6	75.2
Log Source			
Crown Lands	56.8	60.8	67.6
Private Lands	8.0	7.8	7.6
	64.8	68.6	75.2

Source: Price Waterhouse, *The Forest Industry in British Columbia*, 1998.

TABLE 7A
Payments to Governments—By Industry
($ millions)

	1998	1997	1996	1995	1994	1993	1992
Federal							
Current Income Tax	117	95	75	267	395	152	—
Federal Sales Tax	—	—	—	—	—	—	—
Other	16	13	16	13	18	13	9
Total Federal	133	108	91	280	413	165	9
Provincial and Municipal							
Current Income Tax	(17)	23	31	160	410	112	16
Provincial Sales Tax	209	216	208	213	182	180	139
Stumpage, Royalties, Rents, etc.	1,415	1,773	1,733	1,798	1,430	849	616
Lumber Export Tax	26	40	19	—	—	—	—
Property Tax	152	150	152	152	149	154	162
Taxes Included in Electricity Rates	77	83	77	76	75	83	54
Logging Tax	—	6	7	62	124	21	(10)
Corporation Capital Tax	42	45	45	45	44	40	29
Total Provincial and Municipal	1,904	2,336	2,272	2,506	2,414	1,439	1,006
Total	2,037	2,444	2,363	2,786	2,827	1,604	1,015
U.S. Countervailing Duty	—	—	—	(400)	260	140	
Related to Direct Employees	1,648	1,738	1,892	1,873	1,740	1,560	1,421

Source: Price Waterhouse, *The Forest Industry in British Columbia*, 1998.

TABLE 7B

Payments to Government Related to Direct Employees
($ millions)

	1998	1997	1996	1995	1994
Employee Income Tax Deductions	1,329	1,363	1,485	1,479	1,363
Canada Pension Plan	137	156	170	155	145
Employment Insurance	182	219	237	239	232
Total	1,648	1,738	1,892	1,873	1,740
Payments to WCB	196	214	237	199	153

Source: Price Waterhouse, *The Forest Industry in British Columbia,* 1998.

TABLE 8A

Timber Supply Review Results

Area	Pre-TSR AAC	Post-TSR AAC	% Change
CARIBOO REGION			
Timber Supply Area			
Quesnel	2,350,000	2,340,000	-0.4%
Williams Lake	3,975,000	3,807,000	-4.2%
100 Mile House	1,250,000	1,362,000	+9.0%
Tree Farm Licence			
#5 - Weldwood	110,000	110,000	0
#52 - West Fraser	518,952	549,000	+5.8%
Total for region	8,203,000	8,168,000	-0.4%
NELSON REGION			
Timber Supply Area			
Arrow	619,000	619,000	0
Boundary	700,000	700,000	0
Cranbrook	900,947	850,000	-5.7%
Golden	650,000	540,000	-16.9%
Kootenay Lake	900,000	700,000	-22.2%
Revelstoke	269,000	230,000	-14.5%
Invermere	657,000	591,000	-10.0%
Tree Farm Licence			
#3 - Slocan	108,000	65,000	-39.8%
#8 - Pope & Talbot	145,000	145,000	0
#14 - Crestbrook	178,926	164,000	-8.3%
#55 - Evans	110,000	100,000	-9.1%
#56 - City of Revelstoke	110,000	90,000	-9.1%
Total for region	6,048,137	5,484,500	-9.3%
KAMLOOPS REGION			
Timber Supply Area			
Okanagan	2,615,000	2,615,000	0
Merritt	1,254,750	1,254,750	0
Kamloops	2,393,180	2,679,180	+12.0%
Lillooet	643,000	643,000	0

Tree Farm Licence			
#15 - Weyerhaeuser	72,000	72,000	0
#18 - Slocan	187,000	187,000	0
#49 - Riverside	380,000	380,000	0
#33 - Federated Co-op	27,500	22,500	-18.2%
#35 - Weyerhaeuser	125,600	125,600	0
Total for region	7,698,530	8,185,030	+6.3%
PRINCE GEORGE REGION			
Timber Supply Area			
Dawson Creek	1,841,864	1,733,033	-5.9%
Fort Nelson	972,000	1,500,000	+54.3%
Fort St. John	1,803,066	2,015,000	+11.8%
Prince George	9,073,661	9,363,661	+3.2%
Mackenzie	2,974,363	2,977,363	+1.7%
Robson Valley	592,754	602,377	+1.6%
Tree Farm Licence			
#53 - Dunkley	187,630	204,700	+9.1%
#30 - Northwood	407,000	350,000	-14.0%
#42 - Tanizul	132,300	120,000	-9.3%
#48 - CFP	410,000	514,000	+25.4%
Total for region	18,367,638	19,400,134	+5.6%
PRINCE RUPERT REGION			
Timber Supply Area			
Bulkley	895,000	895,000	0
Lakes	1,485,400	1,500,000	+1.0%
Cassiar	140,000	400,000	+185.7%
Kalum	480,000	464,000	-3.3%
Kispiox	1,092,611	1,092,611	0
Morice	1,985,815	1,985,815	0
North Coast	600,000	600,000	0
Tree Farm Licence			
#1 - Skeena	720,000	720,000	0
#41 - West Fraser	430,000	400,000	-23.6%
Total for region	9,188,826	9,317,426	+1.4%

VANCOUVER REGION			
Timber Supply Area			
Arrowsmith	468,500	400,000	-15.3%
Fraser	1,765,000	1,493,000	-12.2%
Kingcome	1,798,270	1,399,000	-22.2%
Mid-Coast	1,000,000	1,000,000	0
Queen Charlotte	509,861	475,000	-6.8%
Soo	580,000	506,000	-12.8%
Strathcona	1,693,745	1,420,000	-16.2%
Sunshine Coast	1,100,000	1,140,000	+3.6%
Tree Farm Licence			
#24 - Western	115,000	115,000	0
#37 - CFP	1,085,000	1,063,000	-2.0%
#38 - Interfor	263,000	263,000	0
#43 - Scott	49,660	44,460	-10.5%
#44 - MacMillan Bloedel	2,680,000	2,228,000	-16.9%
#6 - Western	1,300,000	1,288,000	-0.9%
#10 - Interfor	170,950	170,950	0
#19 - Pacific	978,000	978,000	0
#25 - Western	783,000	779,000	-0.5%
#26 - City of Mission	41,200	45,000	+9.2%
#39 - MacMillan Bloedel	3,675,000	3,700,000	+1.8%
#45 - Interfor	211,000	220,000	+4.8%
#46 - Timber West	558,860	535,000	-4.3%
#47 - Timber West	711,000	865,000	+21.7%
#54 - Interfor	138,000	75,750	-45.1%
Total for region	12,758,670	12,410,160	-2.7%
TOTAL FOR PROVINCE			
TSA	53,362,551	53,430,290	+0.1%
TFL	17,818,578	17,424,960	-2.2%
GRAND TOTALS	71,181,129	70,855,250	-0.5%

Source: BC Ministry of Forests, File 330-04, January, 1997.

Notes

Chapter 1

Much of this chapter is drawn from *The Forests of British Columbia* by Cameron Young (Vancouver: Whitecap Books, 1985). It provides a superb description of the province's forests and is full of excellent photographs. Also helpful was *Ancient Forests of the Pacific Northwest* by Elliott Norse (Washington DC: Island Press, 1990) and *British Columbia: A Natural History* by Richard and Sydney Cannings (Vancouver: Douglas & McIntyre, 1996). Statistical data is from *Forest, Range & Recreation Resource Analysis* 1994 (Victoria: Ministry of Forests, 1995).

Chapter 2

1. H.V. Nelles, *The Politics of Development* (Toronto: Macmillan, 1974), p. 183.
2. *Ibid.*, p. 19.
3. Gifford Pinchot, quoted in Nelles, pp. 185–86.
4. *Ibid.*, p. 186.
5. *Ibid.*, p. 187.
6. *Ibid.*
7. British Columbia Sessional Papers, 1913, p. 65.
8. Letter from H.R. MacMillan, quoted in Ken Drushka, *HR: A Biography of H.R. MacMillan* (Madeira Park, BC: Harbour Publishing, 1995).
9. C.D. Orchard papers, File 8-15, University of British Columbia Library, Special Collections.
10. Gordon Sloan, *The Forest Resources of BC: Report of the Commissioner* (Victoria: King's Printer, 1945), p. 127.

11. *Ibid.*, p. 142.

12. More detailed accounts of this Royal Commission and the events surrounding it can be found in David Mitchell, *W.A.C. Bennett and the Rise of British Columbia* (Vancouver: Douglas & McIntyre, 1983); and Ken Drushka, *HR*.

13. Gordon Sloan, *The Forest Resources of BC*, p. 66.

14. Peter Pearse, *Timber Rights & Forest Policy in BC: Report of the Commissioner* (Victoria, 1976), p. 62.

15. *Ibid.*, p. 235.

16. Ian Mahood and Ken Drushka, *Three Men and A Forester* (Madeira Park, BC: Harbour Publishing, 1990). See Chapter 16 for a lengthy discussion of sympathetic administration.

17. Forest Resources Commission, *The Future of Our Forests* (Victoria, 1991), p. 37.

Chapter 3

1. See *Jack of All Trades*, by J.V. Clyne (Vancouver: Douglas & McIntyre, 1985) and *HR: A Biography of H.R. MacMillan*, by Ken Drushka (Madeira Park BC: Harbour Publishing, 1995).

2. Ken Drushka, *Stumped: The Forest Industry in Transition* (Vancouver: Douglas & McIntyre, 1985), p. 85.

Chapter 4

1. Gordon Sloan, *The Forest Resources of BC: Report of the Commissioner* (Victoria: King's Printer, 1945), p. 127.

2. Data in this chapter is from Price Waterhouse, *The Forest Industry in British Columbia 1998* (Vancouver, June 1999); BC Ministry of Forests, *Forest, Range & Recreation Resource Analysis, 1994* (Victoria, 1995); Council of Forest Industries, *BC Forest Industry Statistical Tables* (April 1994); Vancouver Board of Trade, *The Economic Impact of the Forest Industry on British Columbia and Metropolitan Vancouver* (Vancouver, 1994).

3. Ian Mahood and Ken Drushka, *Three Men and A Forester* (Madeira Park, BC: Harbour Publishing, 1990), p. 172.

4. KPMG; Perin, Thorau and Associates; and Simons, "Financial State of the Forest Industry and Delivered Wood Cost Drivers," report prepared for BC Ministry of Forests, Economics and Trade Branch, April 1997.

Chapter 5

1. C.D. Orchard collection, transcript of interview No. 42, UBC Library, Special Collections, p. 120.

2. Peter Pearse, *Timber Rights & Forest Policy in BC: Report of the Commissioner* (Victoria, 1976), p. 265.

3. *Ibid.*, p. 273.

4. British Columbia, *Ministry of Forests Act*, Section 4 (c), 1978.

5. British Columbia, Commission on Resources and the Environment, *A Provincial Land Use Strategy*, Vol. I (1994), p. 50.

6. *Ibid.*, p. 50.

Chapter 6

1. Andrew Rodgers and Bernhard Eduard Fernow, *A Story of North American Forestry* (Princeton NJ: Princeton University Press, 1951), p. 388.

2. Bernhard E. Fernow, *Economics of Forestry* (New York: Thomas Y. Crowell, 1902), p. 273.

3. *Ibid.*, p. 345.

4. Gordon Sloan, *The Forest Resources of BC: Report of the Commissioner* (Victoria: Queen's Printer, 1957), p. 18.

5. F.D. Mulholland, "Forestry in Sweden and Finland," unpublished report, 1937, p. 3.

6. *Ibid.*, p. 15.

7. "Forest Working Circles," Memorandum to C.D. Orchard, August 27, 1942, Orchard collection, File 5-13, University of British Columbia Library, Special Collections.

8. *Ibid.*

9. *Ibid.*

10. Gordon Sloan, *The Forest Resources of BC: Report of the Commissioner* (Victoria: King's Printer, 1945), p. 78.

11. *Ibid.*

12. *Ibid.*, p. 143.

13. J.D. Gilmour to H.R. MacMillan, September 16, 1946, File 414-2, MacMillan Bloedel collection, University of British Columbia Library, Special Collections.

14. F.D. Mulholland to J.E. Liersch, October 17, 1946, File 414-2, MacMillan Bloedel collection, University of British Columbia Library, Special Collections.

15. H.R. MacMillan to H.J. Welch, MLA, November 27, 1946, H.R.

MacMillan personal papers, University of British Columbia Library, Special Collections.

16. C.D. Orchard, "The Function of the State in the Management of Crown and Private Forests for the Production of an Assured Supply of Wood for Industry," paper presented at 1946 annual meeting of the Canadian Society of Forest Engineers, reprinted in *The Forestry Chronicle*.

17. F.D. Mulholland, "The Relation Between State and Private Forestry," paper presented at 1947 meeting of Canadian Society of Forest Engineers, reprinted in *The Forestry Chronicle*.

18. H.R. MacMillan to J.N. Burke, March 7, 1947, H.R. MacMillan personal papers, University of British Columbia Library, Special Collections.

19. "Some more detail from diaries," Orchard collection, University of British Columbia Library, Special Collections.

20. Sam Dumaresq, *Our Story* (Courtenay: Split Cedar Press, 1996), p. 59.

21. "Transcript of Recorded Interview No. 42," p. 98, C.D. Orchard collection, University of British Columbia Library, Special Collections.

22. Interview by author with Bert Hoffmeister, retired president and chairman of MacMillan Bloedel, March 22, 1993.

23. As has been ably recorded elsewhere: Ian Mahood and Ken Drushka, *Three Men and a Forester* (Madeira Park BC: Harbour Publishing, 1990); David Mitchell, *W.A.C. Bennett and the Rise of British Columbia* (Vancouver: Douglas & McIntyre, 1983); Ken Drushka, *HR: A Biography of H.R. MacMillan* (Madeira Park BC: Harbour Publishing, 1995); Joe Garner, *Never Under the Table* (Nanaimo BC: Cinnabar, 1991).

24. Quoted in Mahood and Drushka, *Three Men and a Forester*, p. 168.

25. *Ibid.*, p. 170.

26. Peter Pearse, *Timber Rights & Forest Policy in BC: Report of the Commissioner* (Victoria, 1976), p. 77.

27. Daowei Zhang and Peter Pearse, "Differences in Silvicultural Investment Under Various Types of Forest Tenure in British Columbia," unpublished paper, November 8, 1994.

Chapter 7

1. Peter Pearse, *Crown Charges For Early Timber Rights*, report of task force on Crown timber disposal, BC Forest Service, 1974.

2. Royal Commission of Inquiry on Timber and Forestry, *Final Report* (Victoria, 1910), p. D31.

3. Gordon Sloan, *The Forest Resources of BC: Report of the Commissioner* (Victoria: Queen's Printer, 1957), p. 513.

4. *Ibid.*, p. 523.

5. British Columbia Ministry of Forests *Annual Report*, various years, compiled by Ray Travers.

6. British Columbia Ministry of Forests *Annual Report*, 1982–1983.

7. *Ibid.* See Tables 7A and 7B.

8. H.V. Nelles, *The Politics of Development* (Toronto: Macmillan, 1974), Chapter 2.

9. H.R. MacMillan to J.D. McCormack, March 30, 1935, H.R. MacMillan personal papers, University of British Columbia Library, Special Collections.

10. *Truck Logger*, August 1962, p. 6.

11. For a full account of this incident see Ian Mahood and Ken Drushka, *Three Men and a Forester* (Madeira Park BC: Harbour Publishing, 1990).

12. Coalition for Fair Lumber Imports, *Lumber Fact Book: The facts, issues and policies behind the Canada/US lumber trade problems* (Washington, DC, June 1986).

13. *Equity*, December 1985, p. 10.

14. "Petter says lumber deal averts trade war" (press release), BC Ministry of Forests, February 16, 1996.

15. Terence Corcoran, *The Globe and Mail*, February 20, 1996, p. B14.

16. IWA Locals 1-217 and 1-357, *The Impact of Log Exports on Employment Opportunities* (Vancouver, 1985).

Chapter 8

1. F.D. Mulholland, "Forestry in Finland and Sweden," unpublished report, 1937, p. 15.

2. Gordon Sloan, *The Forest Resources of BC: Report of the Commissioner* (Victoria: Queen's Printer, 1957), p. 237.

3. "Transcript of Recorded Interview No. 42," C.D. Orchard collection, University of British Columbia Library, Special Collections.

4. Peter Pearse, *Timber Rights & Forest Policy in BC: Report of the Commissioner* (Victoria, 1976), p. 229.

5. *Ibid.*, p. 233.

6. *Ibid.*, p. 235.

7. Memorandum John Cuthbert to W.C. Cheston, File 700-9-6, April 2, 1991.

8. Forest Resources Commission, *The Future of Our Forests* (Victoria, 1991), p. 75.

9. *Ibid.*, p. 81.

10. British Columbia Ministry of Forests, *Timber Supply Review Project, Questions and Answers* (Victoria, May 1992), p. 3.

11. British Columbia, *Forest Act*, Section 7 (3-a-vi).

12. BC Ministry of Forests, News release 1995:107, October 1995.

13. BC Ministry of Forests, News release 1994:069, April 1994.

14. BC Ministry of Forests, News release 1996:072, July 1996.

15. *Forest, Range & Recreation Resource Analysis* 1994, p. 280.

16. BC Ministry of Forests, News release 1997:019, April 1997.

17. *Forest, Range & Recreation Resource Analysis* 1994, p. 95.

Chapter 9

1. The classic work on this subject is *A Brief History of Forestry*, by Bernhard Fernow (Toronto: University of Toronto Press, 1911).

2. For a thorough discussion of the clearcut issue see Patrick Moore, *Pacific Spirit: The Forest Reborn* (West Vancouver: Terra Bella Publishers, 1995); Herb Hammond, *Seeing the Forest Among the Trees* (Vancouver: Polestar, 1991); Chris Maser, *The Redesigned Forest* (Toronto: Stoddart, 1990); Hamish Kimmins, *Balancing Act* (Vancouver: UBC Press, 1997). Although they have diverse opinions on forest management, none of these authors advocate an end to clearcutting.

3. D.M. Smith, *The Practice of Silviculture* (New York: John Wiley & Sons, 1962), p. 2.

4. For a history of reforestation in BC, see Ray Williston, "A Look Back at the Beginnings of the Reforestation Program in British Columbia," in *Learning from the Past, Looking to the Future*, FRDA Report #030, (Victoria: Ministry of Forests, 1988), p. 2.

5. C.D. Orchard, memoirs, p. 103, University of British Columbia Library, Special Collections.

6. British Columbia Ministry of Forests, *British Columbia's Forests: Monocultures or Mixed Forests?* (Victoria, May 1992).

7. For a typical account of these optimistic projections see John Small, "Intensive Management Boosts Yield 300 Percent," *Hiballer*, March 1979, p. 81.

8. Robert Curtis et al, "Intensive Forest Management of Coastal Douglas Fir," in *Loggers Handbook*, Vol. 33 (Washington: US Department of Agriculture, 1973).

9. British Columbia Ministry of Forests *Annual Report*, 1993–1994, p. 79.

10. Forest Renewal BC, *Business Plan, 1996–97* (Victoria), p. 54.

11. Daowei Zhang and Peter Pearse, "Differences in Silvicultural Invest-

ment Under Various Types of Forest Tenure in British Columbia," unpublished paper, November 8, 1994.

12. British Columbia Ministry of Forests, *Silviculture Prescription Guidebook* (Victoria, 1995), p. 1.

Chapter 10

1. Gordon Sloan, *The Forest Resources of BC: Report of the Commissioner* (Victoria: King's Printer, 1945), p. 7.

2. *Ibid.*, p. 84.

3. For a more detailed description of the evolution of multiple-use regulations, see Peter Pearse, *Timber Rights & Forest Policy in BC: Report of the Commissioner* (Victoria, 1976), Chapter 19.

4. *Ibid.*, p. 259.

5. British Columbia, *Forest Act*, Section 28(d)i.

6. *Ibid.*

7. Quoted in *Forest Planning Canada*, Vol. 6, No. 4, p. 45.

8. *Ibid.*, p. 4.

9. *Forest Planning Canada*, Vol. 6, No. 6, p. 10.

10. Forest Resources Commission, *The Future of Our Forests* (Victoria, 1991), p. 88.

Chapter 11

1. *Forest Resources Assessment 1990*, FAO forestry paper #124, Rome, 1995, pp. 17–21.

Index

A

B

C

D